Metallurgy: Techniques and Applications

Metallurgy: Techniques and Applications

Patricia Coleman

MURPHY & MOORE
www.murphy-moorepublishing.com

Metallurgy: Techniques and Applications
Patricia Coleman
ISBN: 978-1-63987-363-0 (Hardback)

MURPHY & MOORE

Published by Murphy & Moore Publishing,
1 Rockefeller Plaza,
New York City, NY 10020, USA

Cataloging-in-Publication Data

Metallurgy : techniques and applications / Patricia Coleman.
p. cm.
Includes bibliographical references and index.
ISBN 978-1-63987-363-0
1. Metallurgy. 2. Metals. 3. Materials science. 4. Chemical engineering.
I. Coleman, Patricia.
TN665 .M48 2022
669--dc23

For more information regarding Murphy & Moore Publishing and its products, please visit the publisher's website www.murphy-moorepublishing.com

Table of Contents

Preface

Metallurgy is a field of material science and engineering which deals with the study of physical and chemical attributes of metallic elements and their inter-metallic compounds and mixtures known as alloys. The basic objective of metallurgy is to separate metals from their ores. The field of metallurgy is further divided into physical, chemical, ferrous and non-ferrous metallurgy. Physical metallurgy studies the physical properties of metals such as tensile strength, yield strength, coercivity, etc. Chemical metallurgy deals with the oxidation and reduction reactions involved in the process of separating metal from its ore. Ferrous metallurgy focuses on those metals which contain iron such as steel while the non-ferrous metallurgy is concerned with the processes involving metals such as aluminium, lead, brass, copper etc. This book attempts to understand the multiple branches that fall under the discipline of metallurgy and how such concepts have practical applications. This book is compiled in such a manner, that it will provide in-depth knowledge about the theory and practice of metallurgy. This book is a complete source of knowledge on the present status of this important field.

A foreword of all chapters of the book is provided below:

Chapter 1- Metallurgy can be defined as the art and science of extracting metals from their ores and modifying the metals for further use. Mechanical metallurgy, physical metallurgy, powder metallurgy, chemical metallurgy, extractive metallurgy, etc. are some of its types. This chapter delves into the topic of metallurgy and its different types.; **Chapter 2**, Metalworking is the process of shaping and reshaping metals or metal objects to create useful objects, parts or large scale structures. Metalworking has three distinct processes – forming, cutting and joining. These processes are performed with the help of metalworking machinery and fluids. This chapter will provide the reader with an in-depth understanding of the subject matter.; **Chapter 3**, Concepts like stoichiometry, thermochemistry, material balance, mineral beneficiation, etc. all fall under the umbrella of metallurgy. This chapter addresses such metallurgical terms and discusses them in detail for an extensive understanding of the subject.; **Chapter 4**, The extraction of metals from their respective ores and their further refinement can be undertaken through various methods. Bayer process is one such method through which Alumina is produced from bauxite. Bessemer process, Betts electrolyte process, patio process, Pidgeon process, Hoopes process and Dow process are some other metallurgical processes which are closely examined in this chapter.; **Chapter 5**, Metallurgical techniques have been developed and improved over time to increase yield and efficiency while reducing costs and material losses. A diverse assortment of processes are used in metallurgy today, a few of them being mining, casting, welding, annealing, calcination, roasting, molding, electroplating, etc. This chapter has been carefully written to provide a better perspective of this topic.

At the end, I would like to thank all the people associated with this book devoting their precious time and providing their valuable contributions to this book. I would also like to express my gratitude to my fellow colleagues who encouraged me throughout the process.

Patricia Coleman

1

Metallurgy and its Types

Metallurgy can be defined as the art and science of extracting metals from their ores and modifying the metals for further use. Mechanical metallurgy, physical metallurgy, powder metallurgy, chemical metallurgy, extractive metallurgy, etc. are some of its types. This chapter delves into the topic of metallurgy and its different types.

Metallurgy is a domain of materials science that studies the physical and chemical behavior of metallic elements, their intermetallic compounds, and their mixtures, which are called alloys. Metallurgy is distinguished from the craft of metalworking. The science and technology of metals and alloys. Process metallurgy is concerned with the extraction of metals from their ores and with refining of metals, physical metallurgy, with the physical and mechanical properties of metals as affected by composition, processing, and environmental conditions and mechanical metallurgy, with the response of metals to applied forces.

Metal Casting.

In its general, modern sense, metallurgy is the science that studies the chemical and physical properties of metals, including how they perform when used for culturally useful industrial purposes. The term often refers to the procedures used in extracting metals from ore, as well as to the processes related to metals purification and alloy production. Copper, tin, silver, and meteoric iron, which was used by the Egyptians to make weapons, all underwent some form of metalworking process in various ancient cultures.

The first evidence of a standard metallurgy technology appeared during the Bronze Age, which started around 3,500 BC, when it was discovered that by heating and combining copper and tin, a bronze alloy could be created. The Iron Age began around 1,200 BC when the Hittites discovered how to extract iron from ore and work it to advance their cultural aims.

Aluminum Casting Process.

The word was originally (1593) an alchemist's term for the extraction of metals from minerals: the ending -urgy signifying a process, especially manufacturing: it was in this sense it was used by the 1797. In the late 19th century it was extended to the more general scientific study of metals and alloys and related processes.

The extraction of iron from its ore into a workable metal is much more difficult. This includes the ancient and medieval kingdoms and empires of the Middle East and Near East, ancient Iran, ancient Egypt, ancient Nubia, and Anatolia (Turkey), Ancient Nok, Carthage, the Greeks and Romans of ancient Europe, medieval Europe, ancient and medieval China, ancient and medieval India, ancient and medieval Japan, amongst others.

Casting Mould.

Extractive metallurgy is the practice of removing valuable metals from an ore and refining the extracted raw metals into a purer form. In order to convert a metal oxide or sulfide to a purer metal, the ore must be reduced physically, chemically, or electrolytically. Extractive metallurgists are interested in three primary streams: feed, concentrate (valuable metal oxide/sulfide), and tailings (waste). Concentrating the particles of value in a form supporting separation enables the desired metal to be removed from waste products.

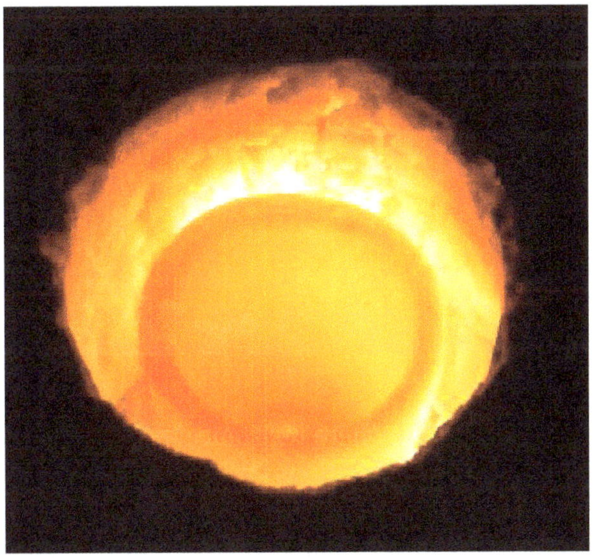

Crucible.

The solution is collected and processed to extract valuable metals. Ore bodies often contain more than one valuable metal. Additionally, a concentrate may contain more than one valuable metal. That concentrate would then be processed to separate the valuable metals into individual constituents.

Hot Rolled Metal.

Common engineering metals include aluminum, chromium, copper, iron, magnesium, nickel, titanium and zinc. Much effort has been placed on understanding the iron-carbon alloy system, which includes steels and cast irons. Cast irons, including ductile iron are also part of the iron-carbon system.

Hot Rolled Metal Coil.

Stainless steel or galvanized steel are used where resistance to corrosion is important. Aluminum alloys and magnesium alloys are used for applications where strength and lightness are required. Copper-nickel alloys (such as Monel) are used in highly corrosive environments and for non-magnetic applications. For extremely high temperatures, single crystal alloys are used to minimize creep.

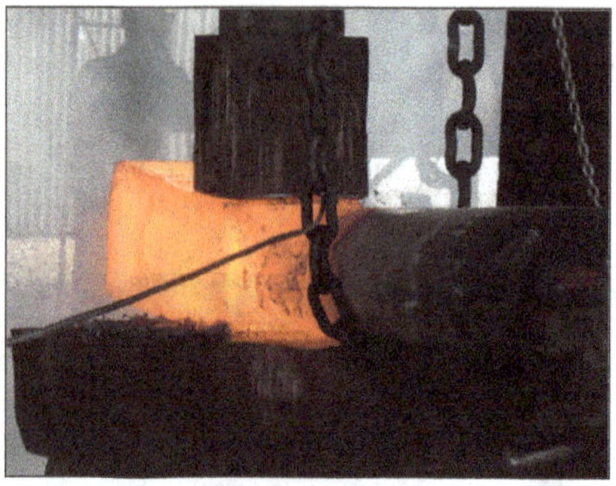

Metal Forging.

This involves the production of alloys, the shaping, the heat treatment and the surface treatment of the product. The task of the metallurgist is to achieve balance between material properties such as cost, weight, strength, toughness, hardness, corrosion, fatigue resistance, and performance in temperature extremes. In a saltwater environment, ferrous metals and some aluminium alloys corrode quickly. Metals under continual cyclic loading can suffer from metal fatigue. Metals under constant stress at elevated temperatures can creep.

Hot Forging Process.

Metals can be heat treated to alter the properties of strength, ductility, toughness, hardness or resistance to corrosion. Common heat treatment processes include annealing, precipitation strengthening, quenching, and tempering. The annealing process softens the metal by heating it and then allowing it to cool very slowly, which gets rid of stresses in the metal and makes the grain structure large and soft-edged so that when the metal is hit or stressed it dents or perhaps bends, rather than breaking, it is also easier to sand, grind, or cut annealed metal.

Underwater Welding.

Quenching is the process of cooling a high-carbon steel very quickly after you have heated it, thus "freezing" the steel's molecules in the very hard martensite form, which makes the metal harder. Tempering relieves stresses in the metal that were caused by the hardening process; tempering makes the metal less hard while making it better able to sustain impacts without breaking. Crystallography allows identification of unknown materials and reveals the crystal structure of the sample.

Process of Metallurgy

The process of extracting a metal in pure form from its ore is known as metallurgy. The process of treatment depends on upon the nature of the ore, impurities, and the metal.

From the extraction point of view, the metals can be broadly classified into five categories:

- Category I: Highly electropositive metals such as alkali metals are extracted by the electrolytic method from their molten/fused salts.
- Category II: Electropositive metals such as alkaline earth metals are extracted by electrolytic method form their molten salt.
- Category III: Reactive metals of high valence are extracted by reduction and electrolytic method.
- Category IV: Heavy metals such as Fe, Co, Ni, Cu, etc. are extracted by roasting and reduction methods.
- Category V: Less reactive metals such as Ag, Au, Pt, etc. are extracted by amalgamation or cyanide process.

The various processes involved in the extraction of pure metals from their ores are:

- Concentration or Benefaction of the ore.
- Extraction of crude metal from the concentrated ore.
- Refining of the crude metal.

Concentration or Benefaction of the Ore

This is a very important part in metallurgy. The ore is obtained from the earth and usually contain dust, stone, quartz, feldspar and other foreign materials. The process of separation of gangue from ore is known as concentration.

Crushing and Grinding of the Ore

Big lumps of ore are first broken into small pieces by the means of mechanical crusher or by hammering. The crushed ore is then ground in a ball mill or a roller mill or a stamp mill. The powdered ore is passed through sieves of the particular size and remains of the sieves are further pulverized in a ball mill or a stamp mill to get ultimately fine dust like powder of theor.

Gravity Separation or Levigation

This method depends on the specific gravity of metallic ore and the earthy impurities. The powdered ore is washed with an upward stream of running water. The lighter gangue particles are washed away while the heavier ore particles settle down and are removed from the base. Generally, oxides like the tin stone (SnO_2), haematite (Fe_2O_3) are concentrated by this method.

Magnetic Separation

If the ore is attracted by the magnet (i.e. ferromagnetic ores, as for example of iron) then it is separated from the non-magnetic impurities by the means of magnetic separator. The mechanism of the magnetic separation is shown in the following figure:

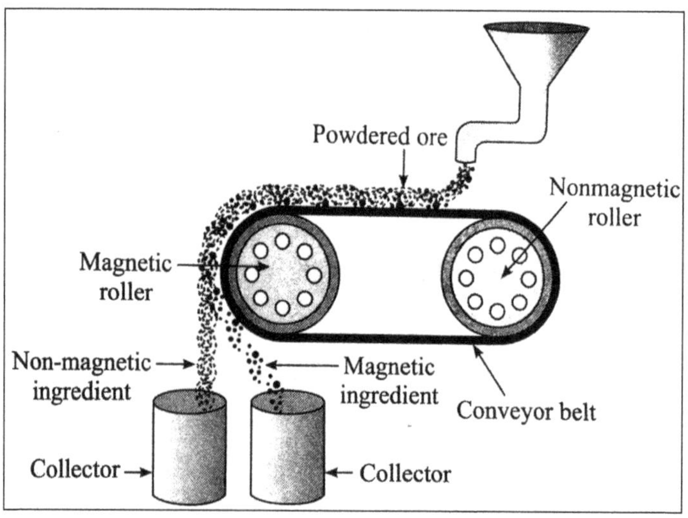

Froth Flotation Process

This is the method based on the principle that different solids have different behavior towards oil and water. Thus certain metal sulphides such as PbS,Zns,$CuFeS_2$, etc. are wetted by certain oils like pine or eucalyptus oil or various organic sulphur compounds and not only by water. In this process, the powder ore is added to the mixture of pine oil and water and agitated by blowing air. The ore which is weighted by oil from froth and come to the surface while impurities settle below. This method is called froth and floatation method as in the figure.

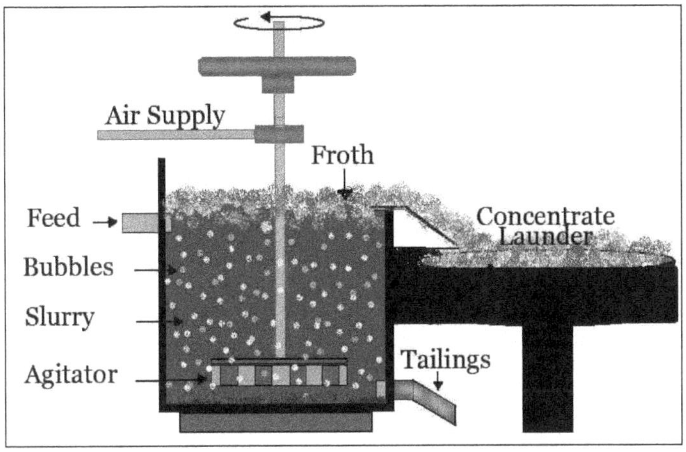

Liquation

This method is used to concentrate ore having the lower melting point than the impurities. For example, stibnite, an ore of antimony which has low melting point, can be concentrated by this method. The powdered ore is heated on the slopping floor of a furnace and the temperature is adjusted to be just above the melting point of the ore. The ore melts and flows while the impurities are left behind.

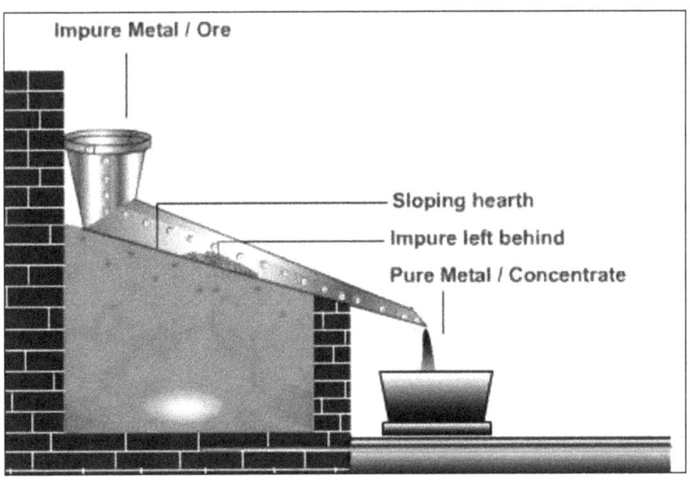

Purification by Liquation.

Leaching

It is a chemical process in which powder ore is treated with suitable reagent. The ore gets dissolve which impurities ore dissolve. This process is called leaching process.

For example – Bauxite is treated with conc. NaOH solution, aluminum oxide is dissolved leaving behind insoluble properties which are removed by filtration.

$$Al_2O_3.2H_2O + 2NaOH \rightarrow 2NaAlO_2 + 3H_2O$$

$$NaAlO_2 + 2H_2O \rightarrow Al(OH)_3 \downarrow + NaOH$$

$$2Al(OH)_3 \rightarrow Al_2O_3 + 3H_2O$$

Extraction of Crude Metal from the Concentrated Ore

The extraction of crude metals from the concentrated ore involves the following steps described below:

Calcination

In this process the concentrated ore is heated in the absence of air to a temperature below the melting points of metal. Calcination is done in a reverberatory furnace. In Calcination,

- Volatile organic matter is driven off.
- Moisture present on the ore is expelled.
- The ore becomes porous.
- Water is removed from the hydrated oxides ores.

$$Al_2O_3.2H_2O \text{ (bauxite)} \rightarrow \text{(clacination) } Al_2O_3 + 2H_2O\uparrow$$

$$Fe_2O_3.3H_2O \text{ (Limonite)} \rightarrow \text{(calcination) } Fe_2O_3 + 3H_2O\uparrow$$

Carbonate ores are decomposed to their oxides,

$$ZnCO_3 \text{ (Zincite)} \rightarrow ZnO + CO\uparrow$$

$$MgCO_3 \text{ (Magnesite)} \rightarrow MgO + CO_2\uparrow$$

$$CaCO_3 \text{ (limestone)} \rightarrow CaO + CO_2\uparrow$$

Roasting

In this process, the ore is heated alone or mixed with other materials under regular supply of air at a temperature below the melting point of the metal. Roasting is usually applied to sulphide ore. During roasting:

- Moisture is driven away.
- Volatile impurities are removed.
- Organic matter is destroyed.
- Sulphur, arsenic and other volatile substance present in the free or combined states are oxidized to volatile oxides.

$$2Zns + 3O_2 \rightarrow 2ZnO + 2SO_2 \uparrow$$

$$4FeS_2 + 11O_2 \rightarrow 2Fe_2O_3 + 8SO_2 \uparrow$$

$$2HgS + 3O_2 \rightarrow 2HgO + 2SO_2 \uparrow$$

Sometimes ore is mixed with suitable material and then subjected to roasting in the absence of air,

$$Ag_2S + 2NaCl \rightarrow 2AgCl + Na_2S$$

$$AgCl + 2Hg \rightarrow AgHg + HgCl$$

Roasting is generally carried out in a reverberatory furnace or in a blast furnace.

Fluxes

The ore even after concentration and calcination still contain certain impurities which have been removed. There are certain substances known as flux which when mixed with calcinated ore combine with impurities to form easily fusible materials known as slag.

Impurity + Flux → Slag

These slags are not soluble. They are lighter than the molten metal and hence float over its surface. These are two types of flux:

- Acidic Flux.
- Basic Flux.

Refining of Crude Metals

The metals obtained by any of these methods described before are seldom pure and hence need further purification. The choice of the purification method depends on the nature of metals and the impurities present. Some purification methods are discussed below.

Oxidative Refining

This is usually where a metal has been prepared by reduction methods because other elements present have also been reduced at the same time. In this method, the impure metal is heated to a high temperature and then exposed to air when the impurities such as carbon, phosphorous and arsenic etc. are oxidized to their volatile oxides which escape along with outgoing gases.

Distillation

This is employed for volatile metals (low boiling point) like zinc (407 °C), mercury (356.58 °C) and cadmium (765 °C) etc. The impure metal is heated in a retort and its vapor are separately condensed in a receiver. The pure metal distils over and the impurities (non-volatile) are left behind in the retort.

Liquation

This method is used for the extraction of metals with low melting points e.g. Tin and Lead. The impure metal is heated on the sloping hearth of a furnace when the molten metal flows away from the infusible impurities.

Electrolytic Refining

This method of purifying the impure metals is employed in the refining of chromium, copper, tin, zinc, silver, lead, gold, aluminum, etc. In this method, the cathode is made of pure metal and impure metal is used as anode as in figure below.

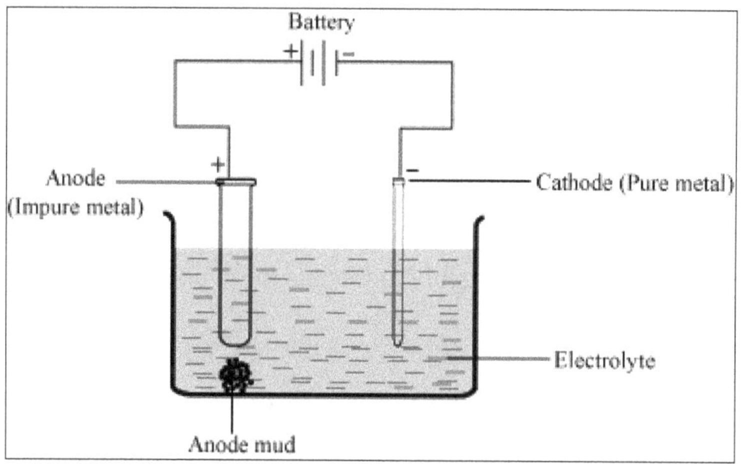

Electricity Refining

Here, copper is refined by this process as:

$$Cu \rightarrow Cu^{+2} + 2e^{-} \ (\text{at anode})$$

$$Cu^{+2} + 2e^{-} \rightarrow Cu \ (\text{at cathode})$$

Mechanical Metallurgy

Mechanical metallurgy can be defined as the interface between an alloy's mechanical behavior, the processing used to produce the alloy, and the underlying structure ranging from the atomic to macroscopic level. Because mechanical metallurgy is most strongly related to how metals break and deform, subfields have developed largely because of disastrous failures such as the Versaille train crash (fatigue), the breakup of the Titanic (toughness), and the sunken Liberty Ships (fracture mechanics). More recently, mechanical metallurgy has evolved in response to challenges in the areas such as energy and transportation.

An excellent example of microstructural engineering to achieve specific mechanical properties is the new generation of steels called advanced high strength steels (AHSS). These alloys, which include dual phase (DP) and transformation induced plasticity (TRIP) steels have combinations of strength and formability that are superior to prior generation steels, so they are being implemented more widely into the automotive industry as car body panels, door beams, and body pillars among many other potential applications. Since these materials are stronger than prior generation steels, they can be used with smaller section thicknesses to allow for vehicle weight reduction and greater fuel economy.

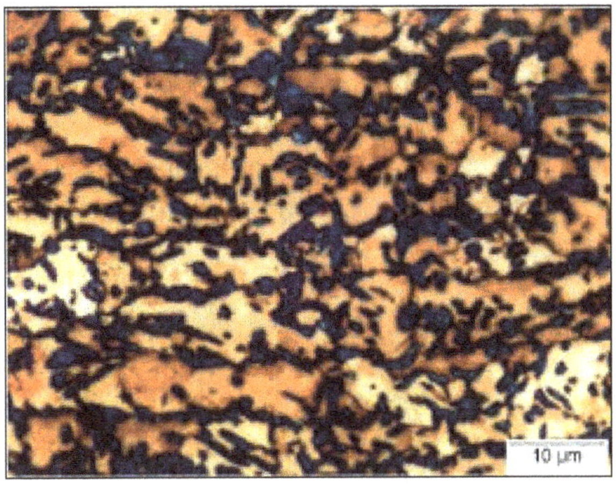

Micrograph, produced by heat tint etching, of TRIP 780 steel. The beige and orange areas are ferrite, the blue areas are martensite, and the dark, blue/purple regions are retained austenite.

DP steels consist of ferrite (relatively soft) and martensite (relatively hard), and the morphology, volume fraction, and composition of each phase can be tailored to achieve different combinations of strength and ductility. TRIP steels contain pro-eutectoid ferrite with varying amounts of retained austenite, martensite, and bainite. The mechanical behavior of TRIP steels is largely dependent on the stability and transformation of retained austenite to martensite during loading. There are many variations of composition, processing, and microstructure of both DP and TRIP steels that a mechanical metallurgist can use to optimize mechanical properties. However, due to the complexity of these steels, there is still much to be understood in areas such as damage evolution, fatigue behavior, and weld mechanical behavior.

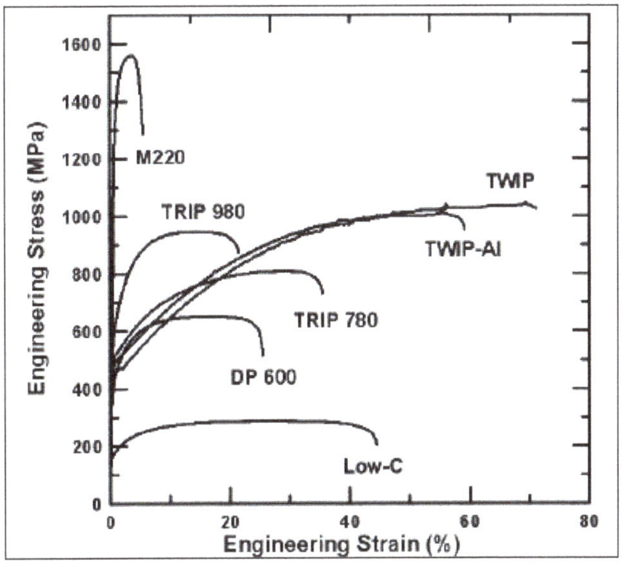

Engineering stress versus engineering strain curves for a low carbon steel compared to several different types of AHSS. TWIP steels rely on mechanical twinning to achieve large amounts of deformation, and M220 is a low ductility, martensitic sheet steel.

Mechanical metallurgy has benefited from the advent of new tools to characterize mechanical response and microstructure. The effects of local grain orientations or overall texture on mechanical behavior can be studied by electron backscatter diffraction (EBSD) imaging. Figure above shows an EBSD image of the inverse problem where a mechanical load imposed during hot deformation influences grain recrystallization behavior in a nickel-base superalloy. Other tools such as thermomechanical processing simulators with advanced control and measuring capabilities, nanoindenters, atomic force microscopes, in-situ mechanical testing devices for electron microscopes, neutron diffractometers[11], and high strain rate testing devices[12]have enabled many advanced mechanical metallurgy studies that have shed light on physical behavior on a microscopic or macroscopic level that could not be achieved until recently.

Figure above shows EBSD inverse pole figure map of partially recrystallized Inconel 945, a nickel base superalloy intended for pipelines exposed to corrosive environments. The small grains are recrystallized grains, and the large grains are deformed but not yet recrystallized. Grain size, texture, and boundary misorientation data can be extracted from EBSD data to help elucidate recrystallization mechanisms during hot deformation that simulates forging practices.

In summary, mechanical metallurgy is a field with expanding opportunities due to the new tools that are available to explore mechanical behavior and also opportunities for advanced materials development as both transportation and energy industries seek to become more efficient.

Origin of Stresses

In equilibrium condition, if there are the external forces acting on the member, there will be some internal forces resisting the action of the external loads.

When you blow up a balloon, after you have tied the end, the air inside exerts a pressure on the surface of the balloon. Imagine the air pushing the surface of the balloon from inside. When you untie the knot and release the air, the balloon deflates and the force on the inner walls is reduced.

What pressure does to the balloon, the stress does to the metal. In that way, pressure and stress are analogous. But the difference is pressure comes into action because you apply force externally, whereas stress is the internal resisting opposing deformation. This also means stresses can be used to quantify the internal resistance.

You apply loads to get stresses. You don't apply a stress. Since pressure and stress are analogous, let us define stress the same way we define pressure i.e. load per unit area. This also implies that the internal force is stress times the area over which it acts.

Important Assumptions

It is very difficult to understand the behaviour of materials if you go about analysing the change at atomic levels. Imagine a glass of hot water kept in front of you. At high temperature the molecules are very active and are vigorously vibrating because of the high energy. Now as you leave the glass to cool in air, the water cools down due to energy loss to surroundings.

You can use a thermometer and get a good estimate of how much the energy decreased (given by the temperature difference). However if you go about trying to measure the change in energy for each molecule, good luck to you.

What we did here is to approximate the change in energy with a change in temperature. Thus, the change is easy to measure and also good enough for most practical purposes. Similarly, for materials we make some assumptions to understand their behaviour. We assume that the material is continuous, that is, it does not have and small voids or empty spaces. We also assume it is homogeneous and isotropic.

When we assume a material is homogeneous, what we mean is if we choose any random section of the material and compare it with its vicinity, the properties are identical. Also, as metals solidify, groups of atoms may be oriented randomly or in one direction. This affects their variation in properties along a said direction. This is known as anisotropy. Again for simplicity, we neglect this and assume the material is isotropic.

Origin of Strains

In a typical experiment in which we wish to study the deformation behavior, we wish to measure the stresses developed in the material as we keep "pulling" it in a given direction.

Let's say you take a metal bar and clamp it at one end. Then you pull the free end by 1mm. Attached to the end is a meter which will give you the force required to pull by 1mm. Then you pull again by 1mm. So now, the displacement is 2mm total. The force reading goes up, because as you pulled to 1mm stresses developed in the sample making it even more difficult to pull further. For every additional pull of 1mm, the force required will keep on increasing because of the development of stress in the material. Also, if you keep pulling the sample, it will fracture and break off.

One way to figure out the happens on loading the sample is to simply plot the force required vs the displacement. But this plot is not a very useful one. Recall that the internal resistance will be a function of area. This means if we have two different samples, one of area A and other of area 2A, we cannot directly compare their force-displacement curves because their areas are not the same.

Stress is calculated as load per area. This means stress is normalised with respect to area. But our original plot was force vs displacement. If we normalise the force, we also need to normalise the displacement. This is where the idea of strains comes into the picture.

Strain is defined as the ratio of change in length to the original length. Let's denote strain by ee. Therefore,

$$e = \frac{\Delta L}{L} = \frac{L\mathit{final} - L\mathit{initial}}{L\mathit{initial}}$$

So, now we have normalised both the force and the displacement from the original experimental setup.

If we have to two different samples, which vary in size and shape and we want to compare these two materials, we will first measure the load-displacement curve. Then we will normalise the force by dividing by the area of the sample and normalise the displacements by converting them to strains. This will result in two normalised plots of stress vs. strain for both the samples we now have a way to compare the two samples.

Looking at the stress-strain curves, we can now start asking questions like, which of these two will fail easily or can this material be used to build a bridge? But for that we will have to get better at reading and interpreting the stress-strain curve.

Powder Metallurgy

Powder metallurgy is a metal-forming process performed by heating compacted metal powders to just below their melting points. Although the process has existed for more than 100 years, over the past quarter century it has become widely recognized as a superior way of producing high-quality parts for a variety of important applications. This success is due to the advantages the process offers over other metal forming technologies such as forging and metal casting, advantages in material utilization, shape complexity, near-net-shape dimensional control, among others. These, in turn, contribute to sustainability, making powder metallurgy a recognized green technology.

The images in include a complex planetary carrier for a four-wheel drive torque transfer system, a helical gear and blades of stainless steel used in laparoscopic surgical scissors, a manifold weighing over 6.5 tons used on an offshore oil platform, and a steel connecting rod used in V-8 engines. All of these components were made using powder metallurgy.

Powder Metallurgy Technologies

In reality, powder metallurgy comprises several different technologies for fabricating semi-dense and fully dense components. The conventional powder metallurgy process, referred to as press-and-sinter, was used to produce the planetary carrier. The surgical scissor parts were formed through the metal injection molding (MIM) process, the manifold was manufactured through hot isostatic pressing (HIP), while the connecting rod was produced using powder forging (PF). Meanwhile, new to the scene, metal additive manufacturing (AM) is gaining popularity.

Using many of these PM processing techniques, as well as other processes such as spray forming, roll compaction, rapid solidification, and others, components are also produced today from particulate materials other than metal powders. Today's advanced materials are seldom made of metals and metallic alloys alone, often incorporating ceramics, ceramic fibers, and intermetallic compounds. These include:

- Cermets.
- Intermetallic compounds.
- Metal matrix composites.
- Nanostructured materials.
- High-speed steels.

Conventional Powder Metallurgy Process

The powder metallurgy (PM) process, depicted in the diagram below, involves mixing elemental or alloy powders, compacting the mixture in a die, and then sintering, or heating, the resultant shapes in an atmosphere-controlled furnace to metallurgically bond the particles.

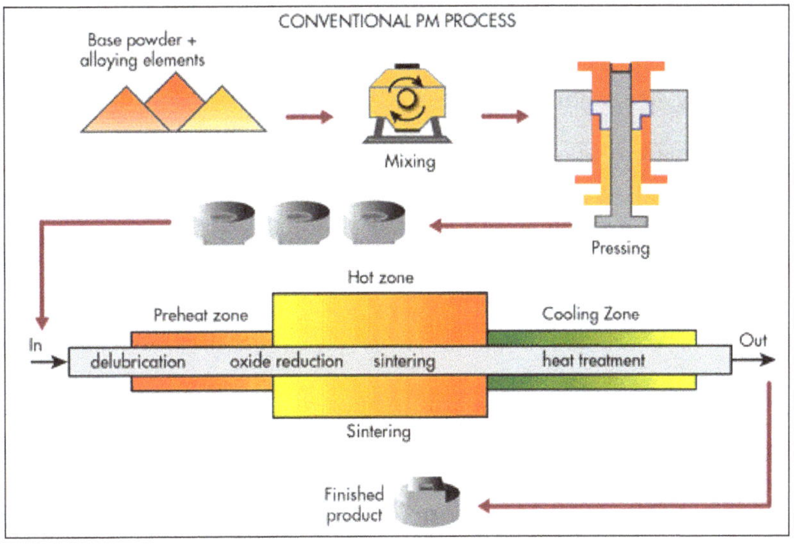

Most powder metallurgy parts weigh less than 5 lb. (2.27 kg), although parts weighing as much as 35 lb. (15.89 kg) can be fabricated in conventional PM equipment. While many of the early PM parts, such as bushings and bearings, were very simple shapes, today's sophisticated PM process produces components with complex contours and multiple levels, and does so quite economically.

Metal Injection Molding

Metal injection molding (MIM) offers a manufacturing capability for producing complex shapes in large quantities. The process utilizes fine metal powders (typically less than 20 microns) that are custom-formulated with a binder (various thermoplastics, waxes, and other materials) into a feedstock. The feedstock is fed into a cavity (or multiple cavities) of a conventional injection-molding machine. After the "green" component is removed, most of the binder is extracted by thermal or solvent processing, and the rest is removed as the component is sintered (solid-state diffused) in a controlled-atmosphere furnace.

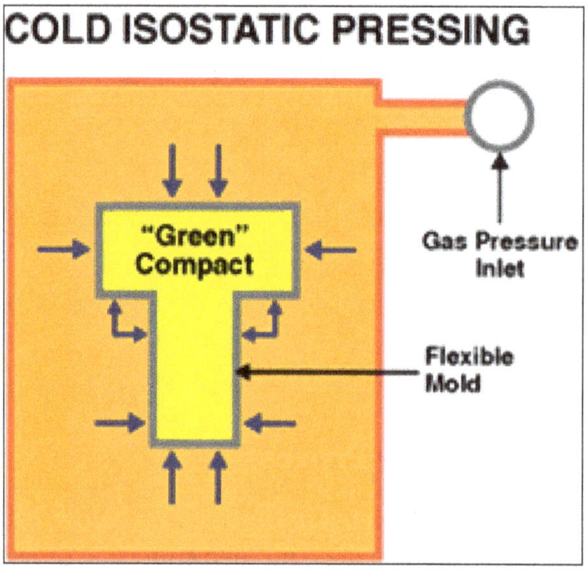

The MIM process is very similar to plastic injection molding and high-pressure die casting, and it can produce many of the same shapes and configuration features. However, it is limited to relatively small (typically less than 250 grams), highly complex parts that otherwise would require extensive finish machining or assembly operations if made by any other metal-forming process.

The advantages of the MIM process lie in its capability to produce mechanical properties nearly equivalent to wrought materials, while being a net-shape process technology with good dimensional tolerance control. Metal-injection-molded parts offer a nearly unlimited shape and geometric-feature capability, with high production rates possible through the use of multi-cavity tooling.

Isostatic Pressing

Isostatic pressing is a PM forming process that applies equal pressure in all directions on a powder compact, thus achieving maximum uniformity of density and microstructure without the geometrical limitations of uniaxial pressing.

Isostatic pressing is performed “cold” or “hot.” Cold isostatic pressing (CIP) is used to compact green parts at ambient temperatures, while hot isostatic pressing (HIP) fully consolidates parts at elevated temperatures via solid-state diffusion. HIP can also be used to eliminate residual porosity from a sintered PM part.

Metal Additive Manufacturing

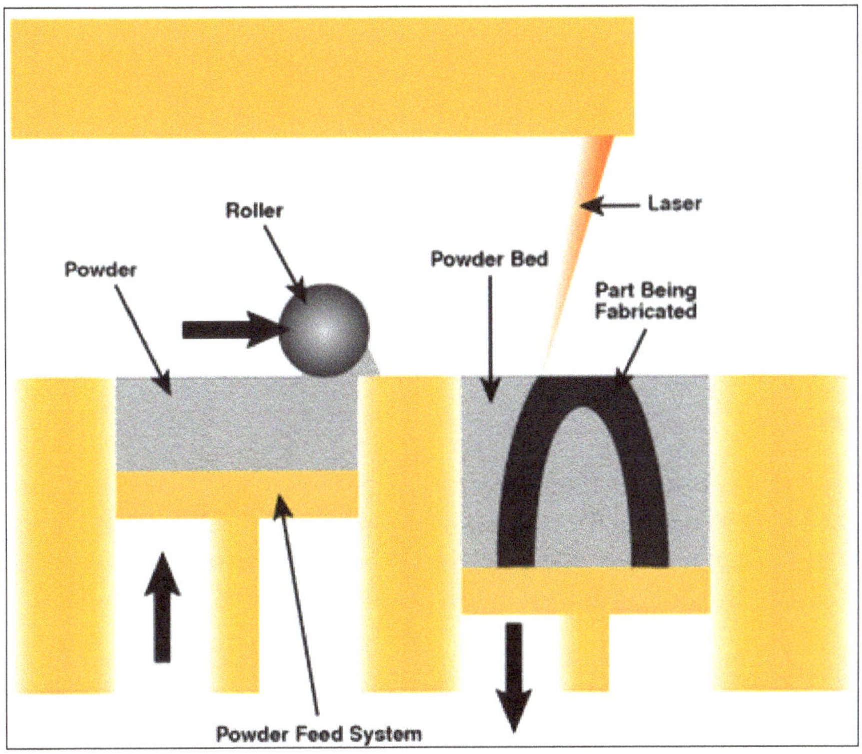

Metal additive manufacturing (AM), or 3D printing, has the potential to profoundly change the production, time-to-market, and simplicity of components and assemblies. Unlike conventional or subtractive manufacturing processes (e.g., drilling), which create a part by removing material, additive manufacturing builds a part using a layer-by-layer process directly from a digital model. It does not use molds or dies that add time, waste material, and expense to the manufacturing process. Additive manufacturing has been used as a design and prototyping tool for decades, but the focus of additive manufacturing is now shifting to the direct production of components, such as medical implants, aircraft engine parts, and jewelry.

Additive manufacturing is not a single type of technology or process. However, while all additive-manufacturing systems employ a common layer-by-layer approach, they use a wide variety of technologies, materials, and processes.

Additive Manufacturing technologies that use metal powders include:

- Laser sintering (LM/SLS/SLFS).
- Electron beam melting (EBM).
- Selective inkjet binding (SIB).

- Laser powder forming (LPF).
- Fused deposition modeling (FDM)/Extrusion.

On describing the technology in these terms, it is not unreasonable therefore to pose the question "Why go to all this effort?"

There are, in fact, many good reasons why Powder Metallurgy might be chosen as the preferred route for the manufacture of a product. In broad terms, these reasons separate into two categories:

- Cost effectiveness: Powder Metallurgy is the most cost effective of a number of possible options for making the part.
- Uniqueness: some characteristic of the product (e.g. combination of chemical constituents, control over microstructure, control over porosity etc.) can be created by starting from a powder feedstock, which would be very difficult or sometimes impossible in conventional processing.

Cost Effectiveness

Product cost effectiveness is by far the predominant reason for choosing Powder Metallurgy and is the main driver of the structural (or mechanical) parts sector. Powder Metallurgy wins the cost competition on the basis of its lower energy consumption, higher material utilization and reduced numbers of process steps, in comparison with other production technologies.

All of these factors, in turn, are dependent on Powder Metallurgy's ability to reduce, or even possibly eliminate entirely, the machining operations that would be applied in conventional manufacture.

In order to eliminate machining operations, Powder Metallurgy relies on its abilities to form complex geometrical shapes directly and to hold close dimensional tolerance control in the sintered product.

VVT System Stator.

Powder Metallurgy's cost effectiveness generally also requires that the particular product be made in large production quantities. If production quantity requirements are too low, there would be no opportunity to amortise the costs of the (long-lasting) forming tooling over a sufficient numbers of parts or to avoid the loss of significant fractions of potential production time in tool changeover/ setting operations.

The production quantities at which Powder Metallurgy would be the process of choice is of course dependent on how difficult it would be to form the shape by a different route, but, in general, would be at least in the order of tens of thousands of parts per year.

Uniqueness

Product uniqueness can be delivered by Powder Metallurgy in a number of different ways:

Processing Combinations of Materials that would Otherwise be Impossible to Mix

Powder Metallurgy allows the processing, in an intimate mixed form, of combinations of materials that would be conventionally regarded as immiscible. Well-established examples of this type of Powder Metallurgy application are:

- Friction materials for brake linings and clutch facings in which a range of non-metallic materials, to impart wear resistance or to control friction levels, are embedded in a copper-based or iron-based matrix.
- Hard metals or cemented carbides, used for cutting tools, forming tools or wear parts. These comprise a hard phase bonded with a metallic phase, a microstructure that can only be generated through liquid phase sintering at a temperature above the melting point of the binder. Tungsten carbide bonded with cobalt is the predominant example of such a material, but other hard metals are available that include a range of other carbides, nitrides, carbonitrides or oxides and metals other than cobalt can be used as the binder (Ni, Ni-Cr, Ni-Co etc.).
- Diamond cutting tool materials, in which fine diamond grit is uniformly dispersed in a metallic matrix. Again, liquid phase sintering is employed in the processing of these materials.
- Electrical contact materials e.g. copper/tungsten, silver/cadmium oxide.

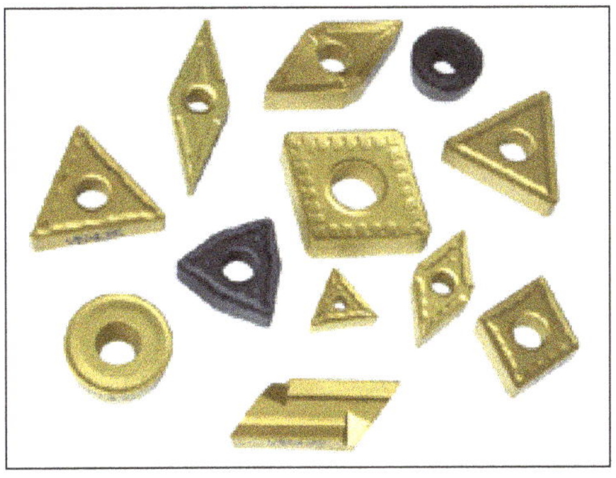

A selection of WC-Co indexable cutting tool inserts.

Processing of Materials with very High Melting Points

Powder Metallurgy enables the processing of materials with very high melting points, including refractory metals such as tungsten, molybdenum and tantalum. Such metals are very difficult to produce by melting and casting and are often very brittle in the cast state. The production of tungsten billet, for subsequent drawing to wire for incandescent lamps, was one of Powder Metallurgy's very early application areas.

Products with Controlled Levels of Porosity

Powder Metallurgy enables the manufacture of products with controlled levels of porosity in their structure. Sintered filter elements are examples of such an application. The other prime example is the oil-retaining or self-lubricating bearing, one of Powder Metallurgy's longest established applications, in which the interconnected porosity in the sintered structure is used to hold a reservoir of oil.

In some specific applications, the generation of superior properties, often through superior control over microstructure, is possible by Powder Metallurgy processing as opposed to conventional casting or wrought routes. Good examples in this category of application are:

- Magnetic materials: Virtually all hard (permanent) magnets and around 30% of soft magnets are processed from powder feedstock.
- High speed steels: The finer and more controlled microstructure from a Powder Metallurgy processed material provides superior toughness and cutting performance than wrought products.
- Nickel- or cobalt-based super alloys: Nickel- or cobalt-based super alloys are used for aero-engine applications, in which Powder Metallurgy processing can deliver compositional ranges and microstructural control not achievable conventionally and therefore an enhancement in operating temperature and performance.

Extractive Metallurgy

The Earth formed from the same cloud of matter that formed the Sun, but the planets acquired different compositions during the formation and evolution of the solar system. In turn, the natural history of the Earth caused parts of this planet to have differing concentrations of the elements. The mass of the Earth is approximately 5.98×1024 kg. In bulk, by mass, it is composed mostly of iron (32.1%), oxygen (30.1%), silicon (15.1%), magnesium (13.9%), sulfur (2.9%), nickel (1.8%), calcium (1.5%), and aluminum (1.4%); with the remaining 1.2% consisting of trace amounts of other elements. Figures illustrates the relative atomic-abundance of the chemical elements in Earth's upper continental crust, which is relatively accessible for measurements and estimation.

Many of the elements shown in the graph are classified into (partially overlapping) categories:

- Rock-forming elements (major elements in green field, and minor elements in light green field).

- Rare earth elements (lanthanides, La-Lu, and Y; labeled in blue).
- Major industrial metals (global production >~3×10⁷ kg/year; labeled in red).
- Precious metals (labeled in purple).
- The nine rarest "metals" (the six platinum group elements and Au, Re, and Te) are in the yellow field.

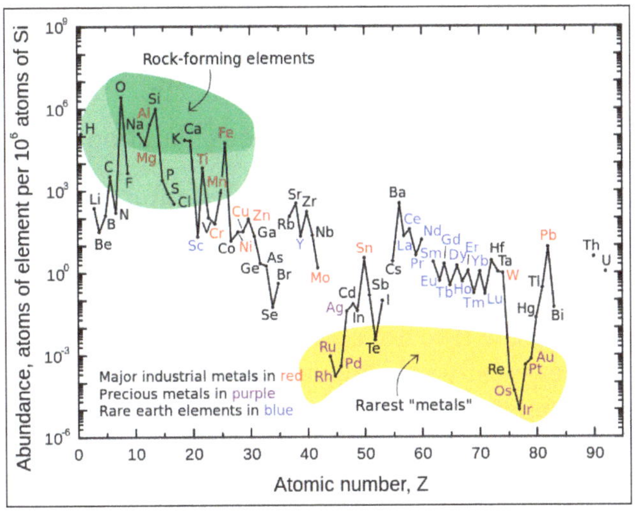

Most metals are found as types of rock in the Earth's crust. These ores contain sufficient minerals with important elements including metals that can be economically extracted from the rock. Metal ores are generally oxides, sulfides, silicates or "native" metals (such as native copper) that are not commonly concentrated in the Earth's crust, or "noble" metals (not usually forming compounds) such as gold. The ores must be processed to extract the metals of interest from the waste rock and from the ore minerals.

Table: Composition of Earth's mantle in weight percent.

Element	Amount	Compound	Amount
O	44.8		
Mg	22.8	SiO_2	46
Si	21.5	MgO	37.8
Fe	5.8	FeO	7.5
Ca	2.3	Al_2O_3	4.2
Al	2.2	CaO	3.2
Na	0.3	Na_2O	0.4
K	0.03	K_2O	0.04
Sum	99.7	Sum	99.1

Extractive metallurgy is a branch of metallurgical engineering wherein process and methods of extraction of metals from their natural mineral deposits are studied. The field is a materials science, covering all aspects of the types of ore, washing, concentration, separation, chemical processes and extraction of pure metal and their alloying to suit various applications, sometimes for direct use as a finished product, but more often in a form that requires further working to achieve the given properties to suit the applications. The field of ferrous and non-ferrous extractive metallurgy has specialties that are generically grouped into the categories of mineral processing, hydrometallurgy, pyrometallurgy, and electrometallurgy based on the process adopted to extract the metal. Several processes are used for extraction of same metal depending on occurrence and chemical requirements.

It takes multiple steps to extract the "important" element from the ore:

- First, the ore must be separated from unwanted rocks.
- Then, the minerals need to be separated out of the ore.
- Since most minerals are not pure metals, further separation methods are required.

Most minerals are chemical compounds that contain metals and other elements.

Concentration

After mining, large pieces of the ore feed are broken through crushing and/or grinding. This step creates particles that are either mostly valuable or mostly waste. TConcentrating the particles of value in a form supporting separation enable the desired metal to be removed from waste products. Froth flotation is one process for separating minerals from the surrounding worthless material by taking advantage of differences in their *hydrophobicity* whereby hydrophobicity differences between valuable minerals other material are increased through the use of surfactants and wetting agents. Flotation is used for the separation of a large range of sulfides, carbonates and oxides prior to further refinement. Phosphates and coal are also upgraded (purified) by flotation technology.

Other ways to concentration minerals include approached based on density (Froth Flotation) used if ore density is less than density of impurities, or via melting points, or magnetic properties (Magnetic separation). This can be used with magnetic ores (i.e. Fe_3O_4) that are passed through a magnetic field, which attracts the magnetic ore and ignore the nonmagnetic impurities.

Diagram of a cylindrical froth flotation cell. The separation of the minerals makes processing complex (that is, mixed) ores economically feasible.

Roasting

Ore is crushed and heated to a high temp using a strong blast of hot air. The process converts the ores to their oxides which can then be reduced. Consider the natural occurring ores of zinc (ZnS (sphalerite) and $ZnCO_3$ (smithsonite). When roasted, smithsonite decomposes to $ZnO_{(s)}$ $ZnO_{(s)}$ and $CO_{2(g)}CO_{2(g)}$ and the hot air involved in roasting sphalerite oxidizes it to produces $ZnO_{(s)}$ $ZnO_{(s)}$ and $SO2_{(g)}SO_{2(g)}$.

$$ZnCO_{3(s)} \xrightarrow{\Delta} ZnO_{(s)} + CO_{2(g)}$$

$$2ZnS_{(s)} + 3O_{2(g)} \xrightarrow{\Delta} 2ZnO_{(s)} + 2SO_{2(g)}$$

Reduction

$C_{(s)}$ and $CO_{(g)}$ is often used as reducing agents in simultaneous reactions. Oxides of Cr, V, and Mn are reduced using Al. During the reduction process the metal oxide is heated to a temperature above its boiling point in order to vaporize it and condense as a liquid. For example, $ZnO_{(s)}$ that is produced in the roasting process is combined with $C_{(s)}/CO_{(g)}$ mixture and heated at a 1100°C. Two reduction reactions are viable at this temperature with either carbon or carbon monoxide acting the reducing agent.

$$ZnO_{(s)} + C_{(s)} \xrightarrow{\Delta} Zn_{(g)} + CO_{(g)}$$

$$ZnO_{(s)} + CO_{(g)} \xrightarrow{\Delta} Zn(g) + CO_{2(g)}$$

Refining

The impurities contained in the metal product of the roasting/reduction process are removed. For example, the Zn(l) produced in the reduction process often contains impurities of Cd and Pb. Though fractional distillation of Zn(l) will work in the refining process, a more commonly used and efficient method is electrolysis. The $ZnO_{(s)}$ produced from roasting is dissolved in $H_2SO_{4(aq)}$ and $Zn_{(s)}$ powder is added allowing the impurities to be displaced. The solution is then electrolyzed and to get a pure metallic $Zn_{(s)}$ as a result.

Zone Refining

A rod containing the desired pure metal and impurities is passed through a series of heating coils and cooled again. The process isolates the impurities that concentrate in the molten zones, leaving the portions behind them somewhat more pure. This process is repeated until impurities are moved to the end of the rod and cut off, resulting in an almost completely pure metal rod.

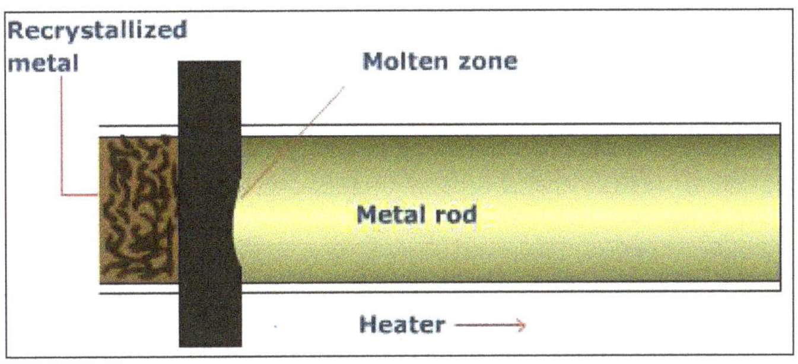

The Zone refining method. from Binod Shrestha.

Thermodynamics of Pyrometallurgy

Pyrometallurgy is the processes of roasting an ore a high temperatures and then reducing its oxide product. Its characteristics include:

- Large amount of waste as a product of concentration.
- High energy consumption to maintain high temps.
- Gaseous emissions that must be controlled (i.e. CO_2 and SO_2).

The oldest and still the most common smelting process for oxide ores involves heating them in the presence of carbon. Originally, charcoal was used, but industrial-scale smelting uses coke, a crude form of carbon prepared by pyrolysis (heating) of coal. The basic reactions are:

$$MO + C \rightleftharpoons M + CO$$

$$MO + \tfrac{1}{2}O_2 \rightleftharpoons M + \tfrac{1}{2}CO_2$$

$$MO + CO \rightleftharpoons M + CO_2$$

Each of these can be regarded as a pair of coupled reactions in which the metal M and the carbon are effectively competing for the oxygen atom. Using reaction $MO + C \rightleftharpoons M + CO < 0$ as an example, it can be broken down into the following two parts:

$$MO \rightleftharpoons M + \tfrac{1}{2}O_2 \Delta G^\circ > 0$$

$$C + \tfrac{1}{2}O_2 \rightleftharpoons CO \Delta G^\circ < 0$$

At ordinary environmental temperatures, reaction $MO \rightleftharpoons M + \tfrac{1}{2}O_2 \Delta G^\circ > 0$ is always spontaneous in the reverse direction (that is why ores form in the first place!), so ΔG° of Reaction $MO \rightleftharpoons M + \tfrac{1}{2}O_2 \Delta G^\circ > 0$ will be positive. ΔG° for reaction $C + \tfrac{1}{2}O_2 \rightleftharpoons CO \Delta G^\circ < 0$ is always negative, but at low temperatures it will not be sufficiently negative to drive $MO \rightleftharpoons M + \tfrac{1}{2}O_2 \Delta G^\circ > 0$.

The smelting process depends on the different ways in which the free energies of reactions like $MO \rightleftharpoons M + \tfrac{1}{2}O_2 \Delta G^\circ > 0$ and $MO \rightleftharpoons M + \tfrac{1}{2}O_2 \Delta G^\circ > 0$ vary with the temperature. This temperature dependence is almost entirely dominated by the $T\Delta S^\circ$ term in the Gibbs function, and thus by the entropy change. The latter depends mainly on Δng, the change in the number of moles of gas in the reaction. Removal of oxygen from the ore is always accompanied by a large increase in the system volume so ΔS for this step is always positive and the reaction becomes more spontaneous at higher temperatures. The temperature dependences of the reactions that take up oxygen vary.

Table: Temperature dependences of the reactions.

Reaction	Δn_g	$d(\Delta G^\circ)/dT$
$C + \tfrac{1}{2} O_2 \rightarrow CO$	0.5	<0
$C + O_2 \rightarrow CO_2$	0	0
$CO + \tfrac{1}{2} O_2 \rightarrow CO_2$	−0.5	>0

A plot of the temperature dependences of the free energies of these reactions, superimposed on similar plots for the oxygen removal reactions $MO \rightleftharpoons M + ½O_2 \Delta G^\circ > 0$ is called an Ellingham diagram. For a given oxide MO to be smelt able, the temperature must be high enough that reaction $MO \rightleftharpoons M + ½O_2 \Delta G^\circ > 0$ falls below that of at least one of the oxygen-consuming reactions. The slopes of the lines on this diagram are determined by the sign of the entropy change.

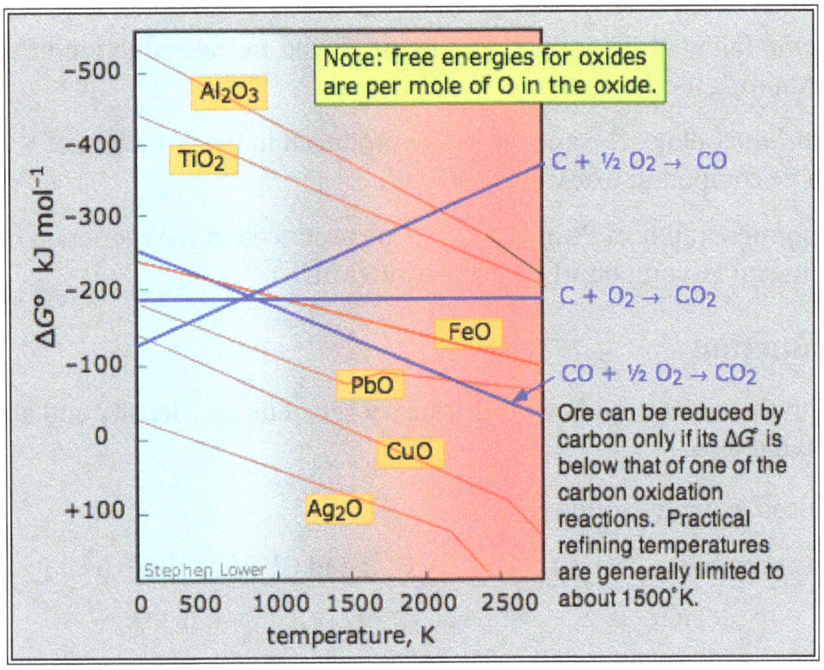

Ellingham diagram. An ore can be reduced by carbon only if its Gibbs free energy of formation falls below that of one of the carbon reduction reactions (blue lines.) Practical refining temperatures are generally limited to about 1500°K.

Examination of the Ellingham diagram shown above illustrates why the metals known to the ancients were mainly those such as copper and lead, which can be obtained by smelting at the relatively low temperatures that were obtainable by the methods available at the time in which a charcoal fire supplied both the heat and the carbon. Thus the Bronze Age preceded the Iron Age; the latter had to await the development of technology capable of producing higher temperatures, such as the blast furnace. Smelting of aluminum oxide by carbon requires temperatures too high to be practical; commercial production of aluminum is accomplished by electrolysis of the molten ore.

In pyrometallurgy, an ore is heated with a reductant to obtain the metal. Theoretically, it should be possible to obtain virtually any metal from its ore by using coke as the reductant. For example, the reduction of calcium metal from $CaO_{(s)}$:

$$CaO(s) + C(s) \underline{\Delta} Ca(l) + CO(g)$$

Unfortunately, many transition metals (e.g., Titanium) react with carbon to form stable binary carbides:

$$Ti + C \rightarrow TiC.$$

Consequently, more expensive reductants (such as hydrogen, aluminum, magnesium, or calcium) must be used to obtain these metals. Many metals that occur naturally as sulfides can be obtained by heating the sulfide in air, as shown for lead in the following equation:

$$PbS(s)+O_2(g)\underrightarrow{\Delta}Pb(l)+SO_2(g)$$

The reaction is driven to completion by the formation of SO_2, a stable gas.

Iron Alloys of Cr, V and Mn

- *Ferrochrome* (alloy of Fe and Cr) can be produced in the reduction of Cr in an Fe compound containing its chromite ore, Fe $(CrO_2)_2$.
- *Ferrovanadium* (alloy of Fe and V) can be produced in the reduction of V that occurs when adding an Fe compound to a V oxide (V_2O_5).
- *Ferromanganese* (alloy of Fe and Mn) can be produced in the reduction of Mn that occurs when adding an Fe compound to a Mn oxide (MnO_2).

Titanium Production

Titanium is used in the military and aircraft industry b/c of its low density and ability to maintain its strength are high temps.

Titanium is produced using the following steps:

- Conversion or rutile or TiO_2 to $TiCl_4$with C(s) and Cl_2 under high heat (> 8000C).

$$TiO_{2(s)}+2C_{(s)}+2Cl_{(2(g)}\underrightarrow{\Delta}Pb(l)\,TiCl_{4(g)}+2CO_{(g)}$$

- Reduction of $TiCl_4$ to Ti with a good reducing agent (i.e., Mg in the Kroll process) carried out at high heat (> 10000C).

$$TiCl_{4(g)}+2Mg_{(l)}\underrightarrow{\Delta}Ti_{(s)}+2MgCl_{2(s)}$$

Titanium was used for 85% of the structure of Lockheed's SR-71B Blackbird.

Both $Ti_{(s)}$ and $MgCl_{2(s)}$ are electrolyzed to produces Cl_2, Mg and titanium sponge (Ti enriched mass).

The mg and $MgCl_2$ are removed via vacuum distillation (reducing of pressure with heat to preferential vaporize volatile metal that are then condenses in a separate container).

- The spong is further treated and allowed with other metal to be sold commercially.

The Kroll process is slow, expensive, and proposes a health/safety risk due to the high temperature vacuum distillation.

Copper Production

Cu ores often contain iron sulfides and to achieve a pure (uncontaminated) Cu product, the ores undergo the following processes:

Step 1: Concentration of ore by flotation.

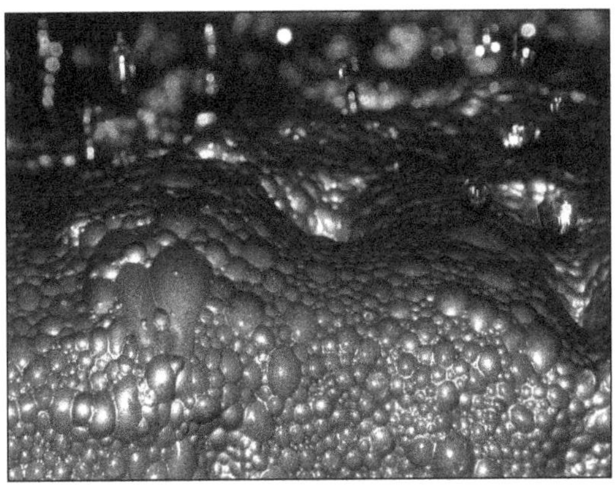

Copper-sulfide foam at the flotation plant of the Prominent Hill mine in South Australia.

Step 2: Conversion of Fe sulfide ores to Fe oxides through roasting (e.g., roasting of Chalcopyrite ore).

$$2CuFeS)2 + 3O)2 \underrightarrow{\Delta} 2FeO + 2CuS + 2SO_2$$

Step 3: Smelting: Mixing with coke/charcoal and heated at high temperatures (i.e. 800°C) for copper to keep Cu from remaining in sulfide form). Impurities are further removed by adding flux if not already present in ore (if gangue is acidic basic flux is added, if gangue is basic acidic flux (i.e., SiO_2) is used. Smelting separates the material to separate into 2 layers (*copper matte*: bottom layer containing molten sulfides of Fe and Cu and *silicate slag*: top layer formed by reaction of oxides of Fe, Ca, and Al w/ SiO_2).

$$FeO_{(s)} + SiO_2(s) \underrightarrow{\Delta} FeSiO_{3(l)}$$

Step 4: Conversion: Air is blown through Cu matte in a separate furnace. This converts the remaining iron sulfide to iron oxide followed by the formation of $FeSiO_3$(l) slag. The slag is poured off and the process is repeated resulting in blister copper with bubbles of SO_2(g) still present.

$$2CuS(l) + 3O_{2(g)} \underrightarrow{\Delta} 2Cu_2O_{(l)} + 2SO_{2(g)}$$

$$2Cu_2O_{(l)} + Cu_2S_{(l)} \underrightarrow{\Delta} 6Cu(l) + SO_{2(g)}$$

Step 5: Blister Cu is further refined through electrolysis to achieve a high-purity Cu product.

Hydrometallurgical Process

The method of refining zinc described above is an example of a hydrometallurgical process and the hydrometallurgical process eliminates the need to control gaseous emissions that are often produced in roasting. Hydrometallurgy is the process of extraction and refining that involves the use of water and aqueous solutions. It is carried out at moderate temps and is generally carried out in three steps:

- Leaching: Metal ions are extracted from their ore by water/acids/bases/salt solutions. Redox reactions that occur are often essential.
- Purification and concentration: Impurities are separated either by absorption on the surface of activated charcoal, ion exchange, or water evaporation; leaving behind a more concentrated solution.
- Precipitation: The process of electrolysis is often used to precipitate the desired metal ions in an ionic solid or reduce them to their free metal.

Extractive metallurgy is the practice of removing valuable metals from an ore and refining the extracted raw metals into purer form. The field of extractive metallurgy encompasses many specialty sub-disciplines, including mineral processing, hydrometallurgy, pyro metallurgy, and electrometallurgy. Especially in hydrometallurgy, the coordination chemistry of the metals involved plays a large role in their solubility and reactivity as the ore is refined into precious metal. Many methods of extractive metallurgy may be applied to get a pure metal from its naturally occurring ore. First ores must be concentrated and separated from their earthly impurities. Concentration of ores may be achieved by froth flotation in which an ore is contained in a froth that floats at the top of a water vat. After the ore has been concentrated it must be roasted to a high temperature allowing for the desired metal to be oxidized. Once the metal is oxidized it is reduced by a reducing agent. Reduction may require the metal oxide to be heated at a high temperature. The metal must be further refined to remove all of the impurities that it may contain. Refining a metal can be accomplished by electrolysis or zone refining. Certain ores such as those of copper and titanium require special treatment before they can be refined. The extraction of iron ore is particularly important for commercial uses because of the mass production of its alloy, steel.

Hydrometallurgy

Hydrometallurgy is part of the field of extractive metallurgy involving the use of aqueous chemistry for the recovery of metals from ores, concentrates, and recycled or residual materials. Hydrometallurgy is typically divided into three general areas:

- Leaching.
- Solution concentration and purification.
- Metal recovery.

Leaching

Leaching involves the use of aqueous solutions containing a lixiviant which is brought into contact with a material containing a valuable metal. The lixiviant in solution may be acidic or basic in nature. The type and concentration of the lixiviant is normally controlled to allow some degree of selectivity for the metal or metals that are to be recovered. In the leaching process, oxidation potential, temperature, and pH of the solution are important parameters, and are often manipulated to optimize dissolution of the desired metal component into the aqueous phase. The three basic leaching techniques are in-situ preaching, heap leaching, and vat leaching.

In-Situ Leaching

In-situ leaching is also called "solution mining." The process initially involves drilling of holes into the ore deposit. Explosives or hydraulic fracturing are used to create open pathways within the deposit for solution to penetrate into. Leaching solution is pumped into the deposit where it makes contact with the ore. The solution is then collected and processed. The Beverley uranium deposit is an example of in-situ leaching.

Heap Leaching

In heap leaching processes, crushed (and sometimes agglomerated) ore is piled in a heap which is lined with an impervious layer. Leach solution is sprayed over the top of the heap, and allowed to percolate downward through the heap. The heap design usually incorporates collection sumps which allow the "pregnant" leach solution (i.e. solution with dissolved valuable metals) to be pumped for further processing.

Dump Leaching

Dump leaching combines characteristics of heap leaching and in-situ leaching. In a dump leach, an impervious layer may or may not be used depending on the dump location. Ore is dumped to allow processing similar to heap leaching, but the physical characteristics of the location allow for a valley or pit to act as the sump.

Vat Leaching

Vat leaching involves contacting material, which has usually undergone size reduction and classification, with leach solution in large tanks or vats. Often the vats are equipped with agitators to keep the solids in suspension in the vats and improve the solid to liquid contact. After vat leaching, the leached solids and pregnant solution are usually separated prior to further processing.

Other Leaching Techniques

In some cases, special leaching processes are required due to refractory nature of the material. These techniques include pressure or autoclave leaching and concentrate leaching.

Solution Concentration and Purification

After leaching, the leach liquor must normally undergo concentration of the metal ions that are to be recovered. Additionally, some undesirable metals may have also been taken into solution during the leach process. The solution is often purified to eliminate the undesirable components.

The processes employed for solution concentration and purification include:

- Precipitation.
- Cementation.
- Solvent Extraction.
- Ion Exchange.

Solvent Extraction

A mixture of an extractant in a diluent is used to extract a metal from one phase to another. In solvent extraction this mixture is often referred to as the "organic" because the main constituent (diluent) is some type of oil.

The PLS (pregnant leach solution) is mixed to emulsification with the stripped organic and allowed to separate. The metal will be exchanged from the PLS to the organic. The resulting streams will be a loaded organic and a raffinate. When dealing with electrowinning, the loaded organic is then mixed to emulsification with a lean electrolyte and allowed to separate. The metal will be exchanged from the organic to the electrolyte. The resulting streams will be a stripped organic and a rich electrolyte. The organic stream is recycled through the solvent extraction process while the aqueous streams cycle through leaching and electrowinning processes respectively.

Ion Exchange

Chelating agents, natural zeolite, activated carbon, resins, and liquid organics impregnated with chelating agents are all used to exchange cations or anions with the solution. Selectivity and recovery are a function of the reagents used and the contaminants present.

Metal Recovery

Metal recovery is the final step in a hydrometallurgical process. Metals suitable for sale as raw materials are often directly produced in the metal recovery step. Sometimes, however, further refining is required if ultra-high purity metals are to be produced. The primary types of metal recovery processes are electrolysis, gaseous reduction, and precipitation.

Electrolysis

Electrowinning and electrorefining respectively involve the recovery and purification of metals using electrodeposition of metals at the cathode, and either metal dissolution or a competing oxidation reaction at the anode.

Precipitation

Precipitation in hydrometallurgy involves the chemical precipitation of either metals and their compounds or of the contaminants from aqueous solutions. Precipitation will proceed when, through reagent addition, evaporation, pH change or temperature manipulation, any given species exceeds its limit of solubility. In order to improve efficiency in downstream processes, seeding to initiate crystallization is often used.

Biohydrometallurgy

Biohydrometallurgy (bio mining or mineral bioprocessing) is a sub-discipline of hydrometallurgy that exploits some attributes of micro-organisms (bio-catalysts) to facilitate and/or enhance the separation of elements from their ores or other materials. Traditional biohydrometallurgy involves the aqueous, inorganic chemistry of acidic sulfate solutions contacted with sulfide concentrates or ores containing valuable metals. As a method of extracting elements, biohydrometallurgy is an alternative to some intense chemical leaching methods. However, its application to the processing of mineral concentrates compares unfavourably with pyrometallurgical processing because it is slower and because, usually, the infrastructure required for pyrometallurgical processing is already in place. Biological processing (of concentrates) would need to demonstrate a greater than 20% advantage over conventional pyrometallurgical processing to interest the mining industry. Similarly, Poulter et al. remarked that industry reluctance to embrace the technology is partly due to the "inherent process and economic advantage of modern smelting technologies" and the "real and perceived risks associated with the introduction of a novel technology". On the other hand, biohydrometallurgy has achieved acceptance for the processing of low-grade, secondary copper sulfide ores on a very large scale, particularly in Chile, and accounts for approximately 20% of annual global copper production. Likewise, pre-treatment of refractory gold concentrates in agitated tanks at commercial scale has a long and successful history from Fairview, South Africa to Runro, the Philippines. Looking to the future, this suite of well-developed and reliable biohydrometallurgical technologies can be applied, for example, to small deposits in remote locations perhaps with unfavourable terrain, to complex ores that are difficult to process, or to concentrates containing impurities that attract smelter penalties.

Man has benefited from the bio-assisted extraction of metals for much longer than the underlying microbiological foundation of the technology has been understood. Accounts from almost 2000 years ago describe the recovery of copper from mine water by cementation (Table 1) but the 'technology' could pre-date these accounts by many centuries, possibly since man started to mine ores and extract metals.

Table: Historical biohydrometallurgy: Approximately 2000 years of anecdotes and evidence up to the discovery of Acidithiobacillus (At.) ferrooxidans.

Chronology	Observation—Report
25–220 A.D.	The Roman Gaius Plinius Secundus (Pliny the Elder, 23–79 A.D.) wrote to the effect that chrysocolla is also artificially produced by slowly passing water through the mine during the winter until the month of June; subsequently the water is evaporated in June and July.
	Notes in old Chinese texts on leaching of rocks and the formation of gall ($CuSO_4$) springs; copper recovery was by cementation on iron.
	The scientist Galen described in situ leaching at a mine in Cyprus.
1086 A.D	The Gall ($CuSO_4$)—copper process described: copper leaching and recovery using cementation on iron as a commercial process; the copper was used for coinage.

1500s	Copper leached from some ores (*in situ*) and mines (Spain, Hungary, Germany) and copper recovered by cementation on iron. A wood cut illustration in Agricola's De re metallica represents the manual collection of copper solutions in wooden 'vats' and transfer to evaporative ponds to concentrate the solutions.
1600s	The extraction of copper from mine water (Peru) described by Alvaro Alonso Barba de Garfias (priest and metallurgist); he was granted a patent to recover copper from mine water.
1800s	Large-scale copper heap leaching (Spain): Heaps of low-grade ore were left for 1–3 years to undergo *natural decomposition*.
1900–1920	Commercial copper production by leaching and copper cementation (Butte, MT, USA); in place (stope) leaching (Cananea, Sonora, Mexico); recovery of copper from underground mine water (Bisbee, AZ, USA).
1921–1940	The oxidation of sulfur by soil microorganisms described;At. *thiooxidans* (previously *Thiobacillus*) isolated and described.
	The concept of the 'biosphere' in which living organisms shaped the environment (biogeochemistry) expounded by Vladimir Vernadskii (1863–1945).
	Biological production of organic acids reported.
	Biooxidation of pyrite first reported and the suggested commercial application of biohydrometallurgy for zinc sulfide.
	Large scale dump leaching of open pit waste rock with recirculation of leach solutions (Bingham Canyon, UT, USA).
1941–1951	Roles of microorganisms in rock-weathering processes recognised.
	Acidithiobacillus ferrooxidans isolated from acid drainage of a coal mine and its role in oxidising Fe(II) and reduced inorganic sulfur compounds (RISC) in acid mine drainage (AMD) described.

Not surprisingly, metal extraction dominated commercial developments before the discovery that microorganisms were involved in the process. However, that discovery created a broad new research field within which the biogeochemical roles of acidophilic microorganisms were explored in relation to acidic drainage at mine sites and geothermal-heated sulfur-rich environments. Naturally, the greater understanding of microbial attributes was incorporated into established or new technologies.

Microbiological Discovery

The discovery and description of *At. ferrooxidans* prompted a rapid increase in microbiological studies related to the oxidation of sulfide minerals. Arguably,*At. ferrooxidans* is the most studiedacidophilic bacterium known, with more than 9000 papers in Web of Science, far more than any other acidophile. However, the perceived importance of *At. ferrooxidans* in bioleaching environments was largely a consequence of the relative ease of its enrichment and isolation, and its rapid growth in acidic iron(II) media, by which it out-competed other members of its microbial community. In the early years (post-1950), source materials of microorganisms tended to be spoils dumps at coal mines or waste dumps at base metals mines. Since then, many acidophiles have been isolated and described and many more are anticipated (Figure below), constrained only by the number of research groups interested in acidophiles. Nevertheless, iron(II)- and sulfur-oxidising acidophiles are relatively few in number, especially those isolated from, or detected in, heap- or agitated-tank bioleaching reactors for the processing of sulfide minerals.

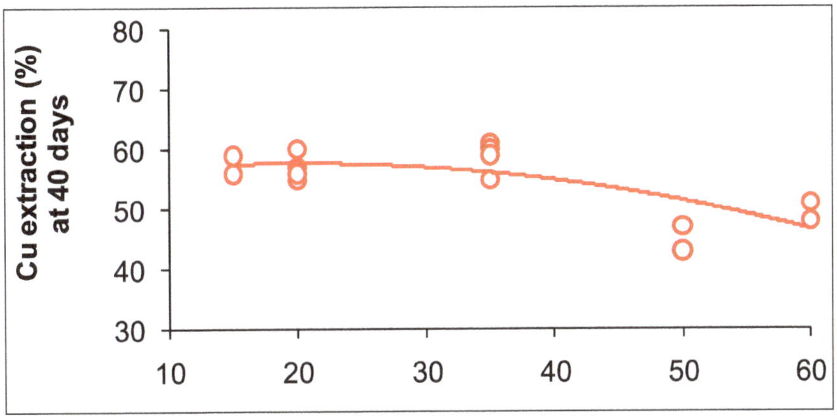

Trend with time of the numbers of acidophilic microorganisms (Bacteria and Archaea) isolated from environments relevant to biohydrometallurgy since the discovery of At. thiooxidans. They are grouped according to whether they are chemolithotrophs or mixotrophs that utilise iron(II) (Fe OX), both iron(II) and sulfur (Fe + S OX), sulfur but not iron(II) (S OX) or are heterotrophs (HET).

More general overviews have been focused on the broader range of microorganisms and microbial interventions associated with acid mine drainage (AMD) environments including collations of acidophilic prokaryotic micro-organisms on the basis of their physiological characteristics. These microorganisms are of direct interest in the context of the microbial communities that can inhabit mineral sulfide heaps. Another group of acidophiles of strong interest are the hyperthermophilic archaea found in sulfur-rich geothermal regions (solfataras, volcanoes, hot springs or combusting coal dumps) that might be used to enhance metal extraction from mineral sulfide concentrates in higher-temperature processes.

Before leaving microbiological discovery, it is worth recapping how biomining organisms assist metals extraction. The microbial capabilities of ferrous ion oxidation and RISC oxidation are the two key functions in the most commonly practised form of biohydrometallurgy, ferric sulfate—sulfuric acid leaching in agitated tanks and heaps/dumps. Microbial ferrous-ion oxidation regenerates the oxidant (ferric ion, Reaction ($4Fe^{2+} + 4H^{+} + O_2$ +microbial catalysts $\rightarrow 4Fe^{3+} + 2H_2O$)) required for sulfide mineral dissolution (e.g., covellite, Reaction ($CuS + Fe^{3+} \rightarrow Cu^{2+} + 2Fe^{2+} + S^0$)). At pH < 4, microbial ferrous ion oxidation is faster than chemical oxidation (e.g., Johnson and references therein). RISC oxidation to sulfuric acid contributes to the acidic environment required for leaching (Reaction ($2S^0 + 3O_2 + 2H_2O$ +microbial catalysts $\rightarrow 2H_2SO_4$)).

$$4Fe^{2+} + 4H^{+} + O_2 + \text{microbial catalysts} \rightarrow 4Fe^{3+} + 2H_2O$$

$$CuS + 2Fe^{3+} \rightarrow Cu^{2+} + 2Fe^{2+} + S^0$$

$$2S^0 + 3O_2 + 2H_2O + \text{microbial catalysts} \rightarrow 2H_2SO_4$$

However, the geochemistry and biochemistry of sulfide mineral dissolution is more complex than is conveyed by Reactions shown above. It encompasses the electrochemical reactions that take place at the mineral surface during sulfide dissolution as well as physiological processes that control the passage of compounds or ions across the microbial cell membrane and within the cell during growth.

Microbial Responses to Extreme and Variable Habitats

The two main processes employed in bio-assisted extraction of metals, heap leaching of low-grade sulfide ores and agitated-tank leaching of sulfide concentrates, present very different habitats for the microbial communities that colonise them. Heaps of low-grade ore particles with a broad size distribution (up to 25 mm) are constructed on a large scale and are essentially dynamic, heterogeneous systems in terms of ore mineralogy and grade. Heat generation is a consequence of microbially-catalysed oxidation of sulfide minerals and heat management requires a suitable balance between heap aeration and irrigation. In contrast, agitated tanks are closer to homogeneous in terms of concentrate grade, solution acidity, provision of O_2 and CO_2 and a selected and controllable temperature for the process. In tanks, there is an abundance of substrate (the sulfide minerals, ferrous ions and sulfur) that diminishes with passage through the train of tanks. Ideally, microorganisms are not exposed to severe fluctuations in conditions.

Much of our understanding of biomining microorganisms, their functions and capabilities, rests upon the collection of acidophiles from mine-impacted sites or geothermal regions that have been isolated and characterised. They form the foundation upon which subsequent studies of the microbial ecology of new sites can be described, monitored and expanded. In particular, the availability of "type strains" deposited in commercial culture collections has facilitated numerous research studies focused on microbial responses to different conditions relevant to bioleaching heap or agitated-tank processes. The four main parameters affecting microbial growth and activity are temperature, acidity, and cations and anions in process water, assuming a substrate is available.

Temperature

Primarily, microorganisms whether or not they are acidophiles tend to be "classified" according to the temperature range within which they grow, for example mesophiles with temperature optima of 20–40 °C; moderate thermophiles (40–60 °C) or extreme thermophiles (60–80 °C). Variable temperature is more likely to impact microorganisms in heaps than in agitated tanks. A number of temperature-related examples of advances in understanding taken from laboratory-, pilot- and commercial-scale studies of columns and/or heaps are reported in the literature. Rapid temperature increases and their impacts on microbial communities can be problematic in heap leaching, especially at start up when the sulfide content is at its greatest. For example, the heat generated during the bioleaching of an ore in a dynamically-controlled, insulated column simulated heat generation in a 5000 tonne test heap of the same ore. The catastrophic effect on the microbial population in that simulation as a consequence of the rapid heating to more than 70 °C is of particular interest. The difference in the rates of heat generation is attributed to the initial cell density in the column (5×10^9 cells per kg ore) being higher than that in the test heap (estimated to be 5×10^5 cells per kg ore).

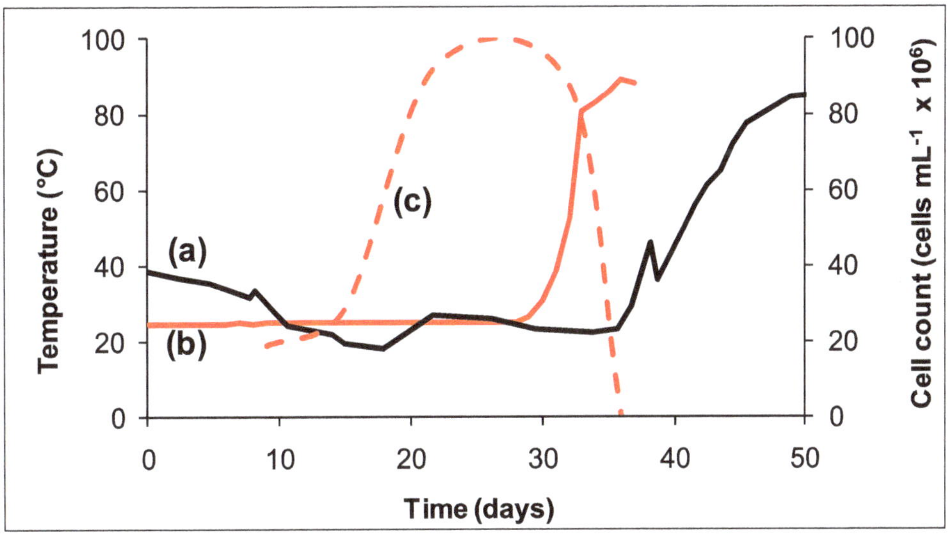

Heat generation profiles (solid lines) for (a) a 5000 tonne test heap of Ni-Cu-FeS ore and (b) the same ore, acid-agglomerated, inoculated and leached in an aerated, dynamically-controlled, insulated column; (c) estimates of cell numbers (broken line) in column effluent representing the microbial response to increased ore temperature.

The determination of preferred temperatures for growth (ranges and optima) of some single strains of acidophiles, based on their iron(II)- or RISC-oxidation rates, illustrates the rapid decline in the activities of microorganisms at temperatures greater than their respective optima. It is generally held that hyperthermophiles, which grow well in sulfur-rich, persistently high-temperature environments such as solfataras and hot springs, would not naturally colonise heaped ores of low-sulfide content. The results from a series of columns subjected to different temperature regimes and inoculated with mesophilic, moderately thermophilic and hyperthermophilic microorganisms as appropriate showed a small benefit from inoculation with hyperthermophiles and led to the subsequent inoculation of a biooxidation heap for refractory gold ore with hyperthermophiles. Recently “patches” of Sulfolobus spp. were found in the extremely large Escondida run-of-mine dump, indicative of the existence of persistent high-temperature regions in the dump, suggesting that thermophiles may colonise large sulfide dumps and heaps that generate and sustain high temperatures over a long period, removing the need for an inoculation strategy.

Acidity

In general, known acidophiles associated with AMD, solfataras and/or managed bioleaching reactors, while widely distributed across the pH range at temperatures favourable to their growth,, ten+d to have a relatively narrow ‘pH window’ below which their activity (iron(II)- or RISC-oxidation) is greatly reduced. However, differences in acid tolerance occur between species/strains. For example, in laboratory flask and reactor tests, a sudden drop in acidity from pH 1.5 to pH 1 caused the cessation of pyrite oxidation by *At. ferrooxidans* but did not affect the activity of a *Sulfobacillus*-like strain. Very few acidophiles prefer or are active in habitats poised below pH 1. This was demonstrated in column tests where solution ORP fell, indicative of lower iron(II)-biooxidation activity, when the recycled, pH 1.7 solution, was acidified further to pH < 1. Similarly, the addition of acid to raffinate prior to recycle to a heap also caused decreased iron(II)- and RISC-oxidising activity in a heap microbial population. Nevertheless, some acidophiles can adapt to very low-pH conditions; *Acidiplasma (Ap.) cupricumulans (formerly Ferroplasma cupricumulans)* grows in the range pH 0.4–1.8, *Ap. aeolicum* in the range pH 0–4 and *Picrophilus (P.) oshimae and P. torridus* can both grow at about pH 0. Acidianus sulfidivorans is another extremely acidophilic archaeon with preferred pH range 0.8–1.4 for growth.

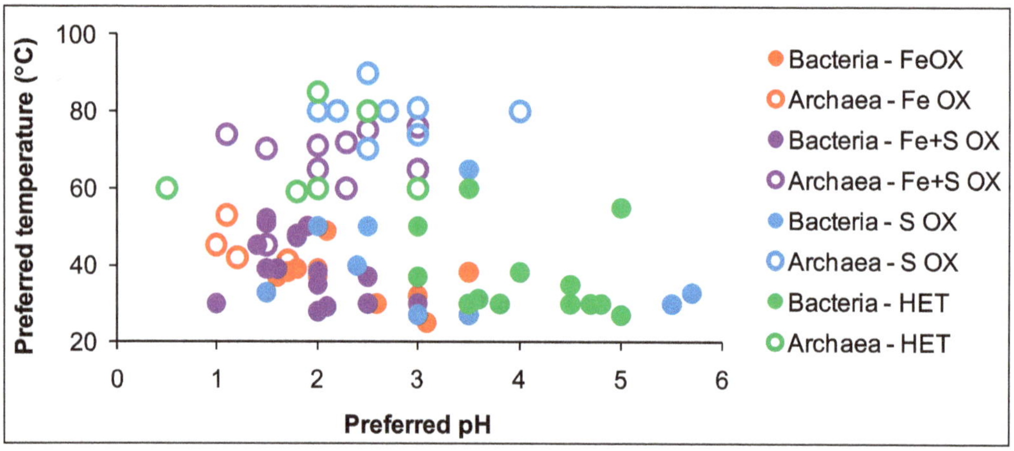

Figure above shows preferred solution pH of microorganisms associated with AMD sites, geothermal regions and managed leaching heap or tank bioreactors, referenced against their preferred growth temperatures. Data obtained from their descriptions or from the recommended growth conditions obtained from commercial culture-collection database.

Cations

Mesophilic and moderately thermophilic microorganisms were adapted by successive culture to different metal cations, up to ($g \cdot L^{-1}$): Co, 15; Cu or Ni, 50; Zn, 98 but, overall, there are limited data on natural adaptation. However, an important point to note is that, while tolerance to particular metals concentrations may increase the viability of a microbial population in a leaching operation, it does not necessarily result in increased metals extraction. Indeed, slightly lower copper extractions were achieved in 40 days by the more Cu-tolerant strains of *S. thermosulfidooxidans* isolated from a commercial copper bioleaching heap.

While data from laboratory studies provide a helpful guide to microbial metals tolerances, more robust data can be obtained from long-term biological processing such as continuous agitated-tank or heap-leach processing. For example, two-years into the continuous pilot-scale operation of a biooxidation plant for the treatment of arsenopyrite gold concentrates, the retention time had reduced from 12 to 3.5 days, the mesophilic microbial population being active in solutions of 13 $g \cdot L^{-1}$ As Microbial adaptation to base metals in pilot- and demonstration agitated-tank plants is less well documented but microbial populations have been developed that tolerate >5 gL^{-1} Co, 23 $g \cdot L^{-1}$ Ni with 38 $g \cdot ^{L-1}$ Fe, or 36 $g \cdot L^{-1}$ Cu in cobalt, nickel and copper continuous plants, respectively.

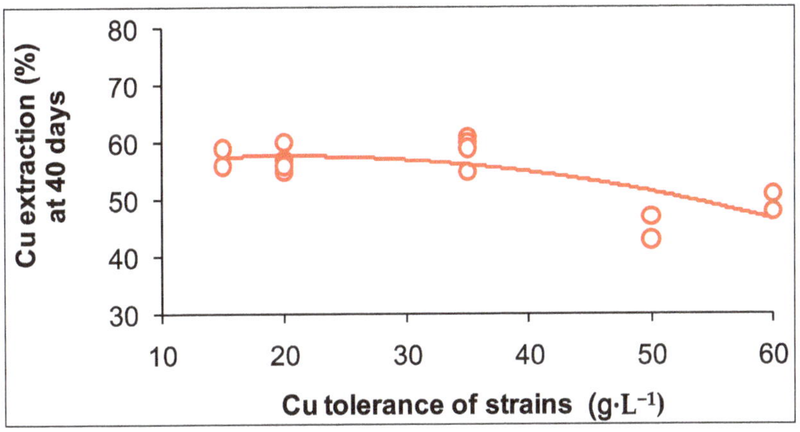

Bioleaching of chalcopyrite concentrate by isolates of *S. thermosulfidooxidans* exhibiting different copper tolerances. Tests were conducted with 3 wt % concentrate in basal salts medium (initial pH 1.8) in a shaking incubator (45 °C, 165 rpm). Initial cell density 5×10^6 cells·mL^{-1}.

From these data it can be inferred that adaptation has taken place, given that the concentrations are much higher than those encountered in most metal-rich acidic environments.

Anions

High concentrations of anions can be present in process waters associated with both heap and agitated-tank bioleaching, but in tank leaching sulfate will dominate. Ore heaps are more likely to contain mineral species enriched in chloride, nitrate or fluoride, which are known to inhibit microbial activity. In bioleaching heaps, especially in Chile, the mineral atacamite ($Cu_2Cl(OH)_3$) presents a problem because of the ease with which it dissolves in acidic solutions and releases chloride ion to the circulating process water. In addition, Chile is an arid region with a scarcity of freshwater, leading to the use of seawater at some mines, a factor that has led to strong interest in the development of bacterial cultures active in seawater media. However, efforts to discover or develop acidophilic iron-oxidising strains or adapted cultures that tolerate chloride concentrations equivalent to seawater have had limited success. A survey of microorganisms from saline and acidic drains, lakes and sediments of Western Australia (52 samples) resulted in only five enrichment cultures that could tolerate between 26–41 g. L^{-1} Cl$^-$ in growth media (pH 2). Cultures that oxidised sulfur generally tolerated higher concentrations of salts than those that oxidised iron(II), consistent with previous data.

Overall, known biomining acidophiles do not tolerate chloride concentrations equivalent to seawater (approximately equivalent to 29 g. L^{-1} chloride salts) regardless of the conditions of growth. In several studies, mesophiles tolerated 2–20 g. L^{-1} NaCl, moderate thermophiles tolerated 10–30 g. L^{-1} NaCl and a mixed culture of thermophilic archaea tolerated up to 15 g. L^{-1} NaCl [70–72,74]. Data from descriptions of bacteria and archaea for salt tolerance referenced against their preferred pH for growth shown in figure below indicated that acidophilic iron-oxidising strains with preferred pH 1–2 have limited tolerance to chloride. Strains with the dual capability of iron- and RISC-oxidation, some within slightly less acidic environments, exhibited greater chloride tolerance, particularly Acidihalobacter prosperus (formerly Thiobacillus prosperus) and some Alicyclobacillus-like strains obtained from marine environments. In Figurebelow, known biomining microorganisms mainly lay below the dotted line (equivalent chloride content to seawater). Halophilic bacteria, including the RISC-oxidising Halothiobacillus spp, and archaea that grow in environments at least 1.5 M NaCl, well above the chloride content of seawater, generally prefer to grow in environments with neutral pH or higher, above the extended acidophile or acid-tolerant range shown in Figure below and of lower acidity than would be useful in bio-assisted metals extraction. The discovery or development through adaptation of halo-tolerant acidophiles with iron(II)- and/or RISC-oxidising capability appears to be a major challenge that must be overcome if seawater is to become widely used in heap (bio)leaching.

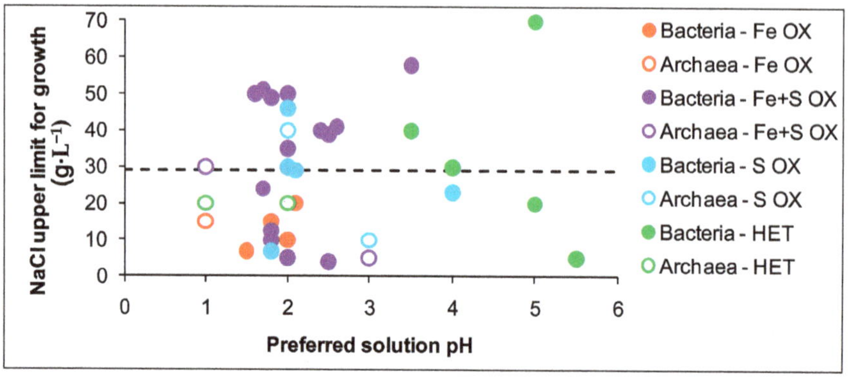

Figure above shows Maximum salt tolerance of acidophiles referenced against their preferred solution pH. The dotted line represents the concentration of chloride salts in seawater. (Data obtained from published descriptions of species or from the recommended growth conditions obtained from the DSMZ-Deutsche Sammlung von Mikroorganismen und Zellkulturen GmbH database).

Concentrations of fluoride higher than 0.5 g.$^{L-1}$ are detrimental to microbial growth, potentially making fluoride in heap process water a greater challenge than chloride. A case in point was a chalcocite heap for which the failure to reach design recoveries was partly attributed to the failure to test the solubility of fluoride-containing minerals in the ore. In that study, concentrations of iron-oxidizing microorganisms were lower than is typical for sulfide heap leaching, diminishing from the critical concentration of 10^5 cells. mL^{-1} to 10^3 cells. mL^{-1} within a year. In an earlier but related study, finely ground particles of a variety of minerals inhibited iron(II) oxidation by At. ferrooxidans but fluorapatite (Ca5(PO4)3F) particles (< 2 wt %).

In a study of sulfur oxidation using At. thiooxidans, the toxicity of fluoride was shown to be pH dependent. It was strongly inhibitory at pH 2.3, where undissociated HF dominates the speciation, and is known to penetrate cell membranes. Fluoride was less inhibitory at pH 4.5, where only 10% is present as dissociated HF, and not toxic at pH 7 where the dissociated MF molecule is dominant (M = Li, Na or K). Fluoride toxicity differs between bacterial species. In a recent comparison of iron(II)- and/or RISC-oxidation by five known bioleaching bacteria, the rank order of fluoride tolerance in media pH 1.5–2 was At. ferrooxidans > At. thiooxidans > Leptospirillum ferrooxidans > At. caldus > Sulfobacillus (S.) thermosulfidooxidans. The fluoride tolerance of bacterial cultures can be increased substantially, in one case from 100 to 850 mg.L^{-1} through adaptive protocols.

The impact of the bioleach solution composition on the concentration of the HF species was investigated at 65 °C using Sulfolobus metallicus growing on pyrite (initial solution pH 1.5) and monitoring changes in ORP as a surrogate measure of microbial activity and growth. The tests varied in the concentration of aluminium added to the growth medium, in the knowledge that Al(III) (and Fe(III)) form Al- or Fe-fluoride complexes. A similar strategy of aluminium addition counteracted fluoride toxicity during the bioleaching of a low-grade, fluoride-containing ore using S. thermosulfidooxidans (50 °C, growth media in the range pH 1.4–2.1), demonstrating that, if fluoride solubilisation is accompanied by strong aluminium or iron(III) solubilisation from gangue minerals, then the effects on the microorganisms may be substantially mitigated due to the formation of aluminium- or iron(III)-fluoro complexes.

As well as being a copper province, Chile is also a nitrate province. The very large Chilean deposits of salitre (saltpetre), mixtures of sodium and potassium nitrate, are a valuable export commodity. It is not surprising therefore that nitrate minerals are sometimes components of sulfide ores and that nitrate in process water may impact on microbial activity. Studies using At.ferrooxidans as the test species showed that nitrate was more inhibitory to iron(II) oxidation than to RISC oxidation. However, in a recent comparison of the effects of nitrate ion on substrate utilisation by bacteria and archaea, it was found that the bacteria adapted to the presence of nitrate in growth media and resumed iron(II)-oxidation within a nine-week period in bioleaching tests. Thus nitrate in ores may have only a transient impact on metals extraction in heaps. In contrast, in the same study, it was found that archaea did not adapt to the presence of nitrate. Based on this result, it was proposed that nitrate may be a useful means of controlling ORP during the higher-temperature oxidation of chalcopyrite in an agitated-tank process.

Surprisingly, the concentrations of sulfate ion, by far the predominant anion in most bioleaching process waters, are seldom reported and the effect of sulfate on microbial activity seems not to be of great concern. Sulfate concentrations can be more than 100 g. L^{-1} in agitated tanks or up to 150 g.L^{-1} in the process water circulating through heaps, concentrations which are much higher than the 40 g.L^{-1} sulfate that caused an adapted mesophilic culture in iron(II) growth medium to halve its replication rate or At.caldus in tetrathionate growth medium to increase the lag time two- to three-fold. During a column bioleaching study (low-grade copper ore), planktonic cells in process water exiting a column declined rapidly by two orders of magnitude when concentrations of ferric and sulfate ions exceeded 30 and 75 g. L^{-1}, respectively , and sulfur-oxidising bioactivity in the Escondida heap was reduced by 50% when sulfate concentrations in process water exceeded 120 g. L^{-1}.

Biodiversity in Bioleaching Environments

The ease with which the microbial diversity of environmental samples can be examined has led to many recent descriptions of the orders, genera and sometimes species inhabiting bioleaching reactors (heaps, dumps, tanks) and/or AMD impacted environments. The collation of example data in figure illustrates the increased knowledge of bioreactor diversity as well as some effects of variable environments on that diversity, for example changes in community structure with time (H1, days 30 to 400; and T1, days 0 to 8), with depth in the heap (H2, depth 1 cf. 3 m) and with operating pH of the process water (H2 and H3, T2). In agitated tanks, the population of attached microorganisms differs from the planktonic community (T3, attached vs. planktonic) and the community structure differs with different solids loading (T3, 10% cf. 20%).

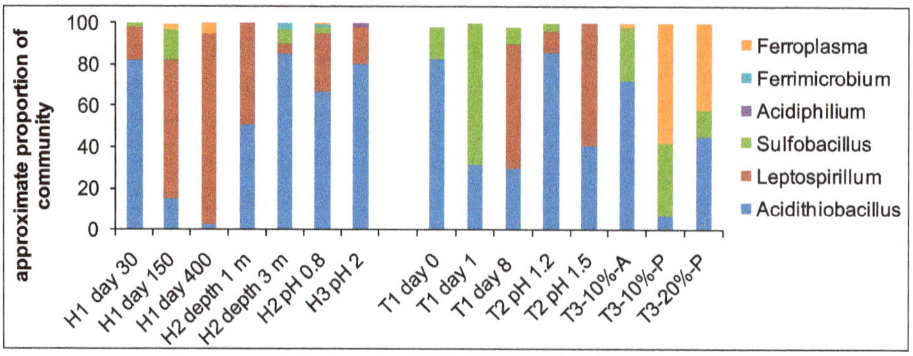

The microbiological diversity to genus level in heaps (H), and agitated tanks (T); A, attached; P, planktonic cells; % solids loading.

In general, abundances of Acidithiobacillus spp. in heaps including At. ferrooxidans and the sulfur-oxidising species At. caldus, At. thiooxidans and At. albertensis are greater early in leaching when there is a greater sulfur content. Leptospirillum and Ferroplasma are more abundant later in the leach when the iron concentrations are higher and possibly inhibitory to Acidithiobacillus and other species. Microbial communities change with time in both heaps and tanks, influenced by parameters such as substrate content and availability, solution composition and acidity, solids loadings in tanks, and temperature. However, comparisons between studies tend to reveal a variety of effects as the communities respond to a combination of environmental parameters and, in many instances, to adapt to them with time.

It is in the nature of sulfide heaps that sulfur contents tend to be low and the heat generated within the bed during sulfide oxidation is transient, essentially because heap-management strategies are employed to maintain temperatures in the range suited to bacterial growth. In general, extreme thermoacidophiles with iron (II) - and sulfur-oxidising capabilities colonise habitats that are persistently hot and sulfurous. As a consequence, it has been assumed that thermophiles such as Acidianus, Metallosphaera and Sulfolobus spp would not naturally colonise sulfide heap operations and that thermophiles would need to be introduced to managed "hot heaps" to exploit increased metal extraction kinetics. However, some recent studies on the biodiversity of sulfide heaps and mine-waste dumps included the putative identification of Archaea and some thermophiles have been detected, including Acidianus, Acidiplasma, Ferroplasma, Metallosphaera, Sulfolobus and Thermoplasma spp.

The advances in both DNA sequencing and protein extraction, accompanied by huge advances in computing power and speed, launched the age of genomics and metagenomics, proteomics and meta-proteomics and metabolomics shown in table below.

Term	Definition
Genomics	The branch of molecular biology concerned with the structure, function, evolution and mapping of genomes (the complete set of genes present in an organism).
Metagenomics	The study of the collected genetic material (genomes) from a mixed community of organisms.
Proteomics	The analysis of the structure, function and interactions of the proteins produced by the genes of a particular cell, tissue or organism; a proteome is the total complement of proteins present in a cell.

Metaproteomics	The study of all protein samples recovered directly from environmental sources; also called community proteomics or environmental proteomics.
Metabolomics	Systematic study of the unique chemical fingerprints that specific cellular processes leave behind; the study of their small-molecule metabolite profiles.

Not surprisingly, the first genome sequence of a bioleaching microorganism was that of At. ferrooxidans. Since then, the number of genomes of microorganisms of direct interest in bioleaching has increased significantly, largely due to a general interest in acidophiles and their activities in AMD environments. The strong overlap with bioleaching is understandable if bioleaching is viewed as an exemplar of extreme but controlled AMD. There were 30 bacterial genomes and 26 archaeal genomes for extremely acidophilic microorganisms, eight metagenome projects of extremely acidic environments (four associated with the AMD at Iron Mountain, CA) and complete sequences for 38 plasmids and 29 viruses from acidic environments. At the same time, it was concluded that these data were not sufficient "to provide a reasonably complete description of the genomic complexity and, by inference, of the full metabolic potential present in bioleaching operations" because (i) there are many gaps regarding the metabolic data; (ii) the microbial diversity of bioleaching and AMD environments is considerably greater than was previously appreciated; and (iii) often genomes were obtained using laboratory-maintained strains that may have accumulated genetic modifications.

Nevertheless, despite the body of research required to address the gaps and the greater diversity in natural microbial communities, the applications of the complementary techniques (In table below) are already augmenting our understanding of the microbial processes involved in extracting metals from minerals. Numerous data on microbial functions and the conditions under which they are exercised best are being gathered. For example, their applications to AMD and bioleaching systems are resulting in better understanding of microbial physiology for individual organisms. Bacterial, archaeal and eukaryotic diversity and their metabolic networks in AMD habitats have been described and the ways in which metabolic activity levels and the partitioning of function are affected by community composition and the physico-chemical environment have been explored. The extent to which each of the individual participants contributes to the process and how they evolve in time to keep the conglomerate healthy and therefore efficient during the entire process of bioleaching is being investigated. Finally, the hypothesis that advanced knowledge of microbial processes can lead to better process design, control and optimization for metals extraction from minerals is supported by the interest shown by some mining companies in transcriptomic analysis of the genomic sequences of newly-isolated (commercially-sensitive) microorganisms with superior ability to extract copper from sulfides.

A valuable output from genomic studies is the generation of genomic sequences for microarray analysis. The development of microarray technology over a number of years has the potential to revolutionize the study of biodiversity in environmental samples. A review on the development of functional gene arrays was introduced as follows: "the recent advance of metagenomic technologies such as high throughput sequencing and functional gene arrays provides powerful high throughput tools for analysing microbial communities". A variety of microarrays have been constructed to contain probes for the genes involved in key microbial functions for specific applications. For example, a microarray was developed based on most of the genes associated with acidophiles, including genes for carbon, nitrogen, sulfur and iron metabolism, DNA replication and repair, and metal resistance. A suite of GeoChip microarrays, of which GeoChip 4 is the most comprehensive , have been used to characterize microbial communities in terms of functional diversity, composition, structure and metabolic activity/capability from a variety of habitats (e.g., soils, aquatic, contaminated sites, bioreactors).

Two modes of microarray application have been described. The use of microarrays based on the complete genomes of the members of a microbial community enables a nearly complete view of gene expression under different bioleaching conditions but it is time-consuming and expensive for the study of organisms isolated from the environment. Examples of specific studies based on complete genome sequences in combination with physiological and biochemical studies include for bacteria Acidithiobacillus spp and for archaea, Metallosphaera sedula. The shotgun DNA microarray is a general approach by which gene expression can be studied in environmental isolates and can be applied to any organism regardless of how much of its genome is sequenced. Applications for acidophiles include bacteria, Leptospirillum spp. and Acidithiobacillus spp. and archaea, Acidianus and Ferroplasma. Applications of microarray analysis of bioleaching and/or AMD environments include the study of bacterial population dynamics at a uranium-contaminated site, the monitoring of acidophile activity in AMD systems and in the Escondida copper heap.

Pyrometallurgy

Pyrometallurgy is a branch of extractive metallurgy. It consists of the thermal treatment of minerals and metallurgical ores and concentrates to bring about physical and chemical transformations in the materials to enable recovery of valuable metals. Pyrometallurgical treatment may produce products able to be sold such as pure metals, or intermediate compounds or alloys, suitable as feed for further processing. Examples of elements extracted by pyrometallurgical processes include the oxides of less reactive elements like iron, copper, zinc, chromium, tin, and manganese.

Pyrometallurgical processes are generally grouped into one or more of the following categories:

- Calcining,
- Roasting,
- Smelting,
- Refining.

Most pyrometallurgical processes require energy input to sustain the temperature at which the process takes place. The energy is usually provided in the form of combustion or from electrical heat. When sufficient material is present in the feed to sustain the process temperature solely by exothermic reaction (i.e. without the addition of fuel or electrical heat), the process is said to be "autogenous" processing of some sulfide ores exploit the exothermicity of their combustion.

Calcination

Calcination is thermal decomposition of a material. Examples include decomposition of hydrates such as ferric hydroxide to ferric oxide and water vapor. The decomposition of calcium carbonate to calcium oxide and carbon dioxide as well as iron carbonate to iron oxide:

$$CaCO_3 \rightarrow CaO + CO_2$$

Calcination processes are carried out in a variety of furnaces, including shaft furnaces, rotary kilns, and fluidized bed reactors.

Roasting

Roasting consists of thermal gas–solid reactions, which can include oxidation, reduction, chlorination, sulfation, and pyrohydrolysis.

The most common example of roasting is the oxidation of metal sulfide ores. The metal sulfide is heated in the presence of air to a temperature that allows the oxygen in the air to react with the sulfide to form sulfur dioxide gas and solid metal oxide. The solid product from roasting is often called "calcine". In oxidizing roasting, if the temperature and gas conditions are such that the sulfide feed is completely oxidized, the process is known as "dead roasting". Sometimes, as in the case of pre-treating reverberatory or electric smelting furnace feed, the roasting process is performed with less than the required amount of oxygen to fully oxidize the feed. In this case, the process is called "partial roasting" because the sulfur is only partially removed. Finally, if the temperature and gas conditions are controlled such that the sulfides in the feed react to form metal sulfates instead of metal oxides; the process is known as "sulfation roasting". Sometimes, temperature and gas conditions can be maintained such that a mixed sulfide feed (for instance a feed containing both copper sulfide and iron sulfide) reacts such that one metal forms a sulfate and the other forms an oxide, the process is known as "selective roasting" or "selective sulfation".

Smelting

Smelting involves thermal reactions in which at least one product is a molten phase. Metal oxides can then be smelted by heating with coke or charcoal (forms of carbon), a reducing agent that liberates the oxygen as carbon dioxideleaving a refined mineral. Concern about the production of carbon dioxideis only a recent worry, following the identification of the enhanced greenhouse effect.

Carbonate ores are also smelted with charcoal, but sometimes need to be calcined first other materials may need to be added as flux, aiding the melting of the oxide ores and assisting in the formation of a slag, as the flux reacts with impurities, such as silicon compounds.

Smelting usually takes place at a temperature above the melting point of the metal, but processes vary considerably according to the ore involved and other matters.

Refining

Refining is the removal of impurities from materials by a thermal process. This covers a wide range of processes, involving different kinds of furnace or other plant. The term "refining" can also refer to certain electrolytic processes. Accordingly, some kinds of pyrometallurgical refining are referred to as "fire refining".

References

- What-is-metallurgy: metallurgyfordummies.com, Retrieved 12, May 2020
- Metal-and-metallurgical-principles/extraction-of-metals-metallurgy, chemistry: kullabs.com, Retrieved 02, August 2020
- Powder-metallurgy-processes, materials-21834821: machinedesign.com, Retrieved 11, January 2020
- Why-powder-metallurgy, introduction-to-powder-metallurgy: pm-review.com, Retrieved 03, July 2020

2

Metalworking: An Introduction

Metalworking is the process of shaping and reshaping metals or metal objects to create useful objects, parts or large scale structures. Metalworking has three distinct processes – forming, cutting and joining. These processes are performed with the help of metalworking machinery and fluids. This chapter will provide the reader with an in-depth understanding of the subject matter.

Metalworking has played a crucial role in human history. Part science, part art and part practical fabrication, metalworking has been an essential driving-force in the development of human society and continues to be central to our capability and development.

Like all industrial processes, metalworking has undergone massive changes throughout our history. However, in many ways, the essential elements of metalworking — forming, casting and joining — remain surprisingly unchanged from the days of our ancestors.

Metalworking is a craft that has allowed secondary processes to develop and was crucial in helping different societies to develop wealth, essential technologies and useful tools throughout history. Societies with advanced metalworkers have always had an edge in terms of wealth and relations with other nations.

Today, we may take metal for granted, but it is important to remember that extracting, processing and working with metal is a very highly advanced process. In fact, it is truly amazing that human beings discovered it so early in our history — and that many civilizations developed these technologies independently from one another.

There is evidence that humans were forging copper as early as 8700 BCE, and civilizations around the world were working with metal by 6000 BCE. Most ancient metalworkers typically worked with iron, tin, lead, copper, mercury, silver and gold, developing different processes for extracting, forming, casting and joining materials.

Economic Importance

Metalworkers have always been very important to how our society functions, but in ancient times, metalworkers were so important they sometimes directly influenced the economic prosperity of a town or village.

Metalworkers often forged a connection between different villages and communities. Their sharing of knowledge and technique helped to form bonds between groups of people that otherwise often had very little in common.

Branches of Metalworking

Three major branches or components make up the art and craft of metalworking.

- Metal Forming: Metal forming is the process of shaping and reshaping metal using mechanical force or heat. The shape and structure of the piece of metal are altered without adding or removing any materials. To help you picture the process, imagine shaping a piece of Play-Dough with your hands. Though easier, it will give you an idea of the process that metal formers go through to create a new item. Metal forming is essential to metalworking, providing the main pieces that make up larger components of the appliances, tools and electronics that we use every day of our lives.
- Metal Casting: Metal casting is an art that has truly helped to make mass production of metal products possible and affordable. Metal casting involves creating a mold that molten metal can be poured into and allowed to dry. The final product is a new and standardized shape of metal that can be specialized in any way needed.
- Metal Joining: Once essential metal parts have been formed and cast, there also needs to be a way to join them together! Metal joining can be done in many ways, and the choice may depend on the metal being used, or the way a product will ultimately be used. Joining might be done through heating and melting metals to act as a solvent to create a bond. In some other cases, metalworkers will use a process called riveting, creating special holes and assembly points to bring parts together — very much in the same way a door hinge functions. Riveting is an ancient process of metal joining that is still sometimes used today, particularly in artisan metalworking.

Many metalworkers specialize in one branch of metalworking, working in connection with other professionals to bring together the best possible finished products.

Today, metalworking continues to play an essential role in our daily lives, and many of the processes we continue to use aren't as different from ancient man's as you might think! As mass production ramped up throughout the 1900s, metal became a household staple in countless ways, being used for everything from appliances to decorations to lunchboxes. With this change came a lot of development for metalworkers, who have found new ways of meeting challenges, better ways of forming and discovered more efficient casting techniques over time.

The Advent of Adhesives

One of the biggest developments in metal joining in the past two decades is the development of new adhesives, capable of joining metal without the need for soldering, brazing or welding. These glues and adhesives have helped to modernize metalworking and make it faster and more efficient.

In some ways, the introduction of metal glue has also helped to narrow the field of metalworkers to specialized work. Now, more assemblers than ever can join metals quickly and easily, without the need for expert soldering, welding or other joining services. What's more, average people have more adhesive options than ever, even when working with metal, which is excellent news for homeowners and those in construction.

Glues and adhesives that allow metals to bond have made the metalworking industry more efficient over time. There are many solutions for small working projects and even glue guns that contain glues for bonding metals of all kinds. This makes metalwork assembly faster, more efficient and more accessible to more people.

Global Metalworking

When our ancestors began working with metal, their access to knowledge, materials and supplies was very limited. Metalworkers were restricted to the local area in which they worked and resided, with limited connection with other parts of the world.

Today, advanced shipping techniques and the possibilities created by globalization have allowed metalworking to become a truly global force, and many of the metal pieces combined to make the appliances we use every day come from all over the world. Although plastic has emerged as one of the most popular and durable manufacturing products in the world, metal and metalworking are still essential to the manufacturing industry.

The future of the metalworking industry is bright. As electronics and home appliance manufacturing develop, metalworkers are challenged to find innovative solutions to creating smaller, more compact and more easily assembled cases, and housing and appliance suppliers. Durability without too much weight or bulk is the demand of every manufacturer, and metalworkers continue to rise to the challenge.

Sheet Metalworking

Sheet metal is not a specific material. Rather, the term is used for metals that have been formed into shape. As a rule, sheet metal has a rectangular shape. It is also characteristic that the thickness of a sheet is considerably below that of its length and width.

In addition to steel, numerous other metals such as copper, aluminum and gold as well as brass alloys can be processed into sheet metal. In order to be processed into sheet metal, however, the materials used should have specific characteristics, i.e. a certain toughness and rigidity, and of course be ductile, i.e. formable. These requirements result in restrictions with regard to the materials that can be used. Thus, it is not possible to process very brittle or extremely stiff materials into sheet metal.

Sheet metal working: Lasertube family from BLM: the LT7 cuts tubes and profiles.

According to this, sheet metal is a semi-finished product that can be produced over a large area, whereby sheets are generally light, thin, stable, elastic and flat. Due to these specifics, they are suitable for all types of cladding and covering. In addition, sheet metal can be deformed in many ways, i.e. it can be welded, bend, punched or cut. A wide variety of shapes can be made from sheet metal, which is why it is used as the starting material for a wide variety of products.

The term "sheet metal processing" covers a large number of manufacturing processes. Basically, the word sheet metal working refers to the manufacture of products, components and parts made of metal. In addition to welding, cutting and bending, the processes of sheet metal processing also include punching, forming, rolling and joining. Sheet metal forming is usually also included in sheet metal processing, with the following subcategories:

- Welding,
- Gluing,
- Bending,
- Piercing,
- Drawing,
- Steel construction,
- Layer forming,
- Heating coil,
- Cutting,
- Slitting,
- Bending punching,
- Laser cutting,
- Laser processing.

Within the framework of sheet metal processing, both soft and resistant sheets can be produced, whereby the different characteristics are achieved by different alloys. For example, there are numerous elements that can be added to steel in the liquid state during metalworking and that determine the material properties of the sheet metal produced. In addition to silicon, nickel and chromium, these elements also include titanium, copper, niobium and molybdenum.

Car construction is a textbook example of sheet metal working.

In metalworking, sheets can be made from different types of steel as well as aluminum and other metals - these are again available in different alloys. In order to categorize the different sheets, there are standards subsuming the materials according to their properties. In addition, material numbers were introduced to ensure a clear structuring of the sheets.

Anyone looking for the oldest form of metalworking will inevitably come across forging. In forging, metals are machined by means of force, for example with the aid of a hammer. Nowadays, this form of metalworking is very rarely found in the industrial sector and is rather used in the artistic sector. The first metal sheets were made of soft materials such as silver and gold. In the course of metalworking, the blacksmiths were able to shape these materials into thin sheets, from which jewelry and coins were subsequently made for instance.

A little bit later, the materials iron and copper also found their way into sheet metal processing. Especially in the High and Late Middle Ages iron sheets were used to make knight's armor. At that time, rolling was still unknown in metalworking. Instead, the sheet metal was processed with hammers or hammer blows until the materials had the desired thickness. Of course, sheet metal and silver and gold were valuable materials - solely because of the value of the raw materials. However, iron sheets also became an expensive commodity, because the sheet metal processing was very complex and time-consuming. The raw materials had to be formed into sheet metal by pure manual work and with the aid of simple tools. Sheet metal processing therefore required a high degree of practice and experience.

When large hammers driven by water power were first used, this led to simplifications in metalworking. Although the hammers shortened the sheet metal working time due to their enormous dimensions and weight, the production time could still take several days, depending on the size of the sheet metal. The first metalworking machines were finally developed during the industrial revolution. These were used to roll metal ingots to produce sheet metal. This not only shortened the production time in metalworking enormously, but also improved the quality of the sheets in terms of flatness and thickness. As a result, sheet metal not only became cheaper, but was also used to manufacture a large number of other products.

Meanwhile, sheet metal processing and metal working have also arrived in modern times and offer a multi-faceted selection of materials, material thicknesses and formats. It should also be noted that the respective sheet metal properties can be adapted to the planned location of use by metalworking.

Manufacturing Process of Sheet Metal Processing

In the following, the different manufacturing processes of metal processing and sheet metal working are presented.

Forming - Rolling, Bending, Forging and Drawing

If one follows DIN 8580, the manufacturing process of forming constitutes a major group in metalworking. With this manufacturing process, the cohesion of the material is maintained. However, in the course of sheet metal processing, the already existing shape of a workpiece is modified by plastic deformation. An example of forming in metal processing is the rolling of steel.

Servo-electric press brake with multi-spindle drive from Coast One: Each of the spindles is driven by its own servo motor. The parallelism between the upper and lower tool is thus set and adjusted in a targeted, precise and fast manner.

In general, plastic deformation occurs in the forming process when the atoms, which are located on the so-called sliding planes, begin to shift as a result of external forces. The modifications of the technological and physical properties of the material occurring during sheet metal processing are subsumed under the term "hardening".

In general, the resistance to deformation of metals at room temperature is higher than at a corresponding forming temperature. Plastic forming of steel, for example, can be achieved at high temperatures with far less work and effort than at room temperature. In addition, high temperatures during sheet metal processing contribute to the immediate recrystallization of the microstructure and do not lead to hardening.

Cold and Hot Forming

In sheet metal forming, a distinction is made between hot and cold forming. Hot forming is a sheet metal forming process in which greater formability is achieved at high rolling temperatures, whereby the occurring forces are lower than in cold forming. The disadvantages of hot forming, however, are its poorer dimensional tolerances and scored surfaces.

In contrast, much more precise dimensional tolerances can be achieved in the cold forming process within the framework of metal processing. In this method of sheet metal working, the material is not heated, which increases its strength. In contrast to cold forming, the quality of the surfaces is also of high quality. A disadvantage, however, is that no welding suitability is ensured after cold forming. The workpiece can only be welded when it has been soft annealed in the course of sheet metal processing, thus minimizing stress.

Separation - Cutting and Punching

Within the framework of the separating manufacturing process, which according to DIN 8580 also forms a main group among the manufacturing processes, the shape of a solid body, i.e. workpiece, is changed by locally eliminating, i.e. reducing, the cohesion. In this form of sheet metal processing, the final shape of the workpiece is therefore contained in its original form. Separation is a technique of metalworking, which also includes the dismantling of composite bodies. The main group "Separation" can be subdivided into six subgroups. There are the following subgroups:

- Splitting.
- Ablation.
- Disassembling.
- Cleaning.
- Machining with geometrically determined cutting edges.
- Machining with geometrically indeterminate cutting edges.

Cutting processes in sheet metal working include sawing, milling, filing, drilling, turning, lapping, thermal cutting, broaching and honing. Wedge cutting and shear cutting, on the other hand, are separating processes. During metalworking, the actual cutting process takes place at the point of action, i.e. at the point where the tool or machine acts on the work piece. With regard to the cutting process in sheet metal working, it should be noted that relative movements, i.e. cutting, feed and infeed movements, are necessary and are carried out by one of the active partners. The power or energy supplied from outside to carry out the separation process is converted into friction, deformation and separation power at the point of action. They are dissipated in the form of heat via the active pair.

Joining - Welding and Soldering

Joining, too, is one of the main production groups in sheet metal processing manufacturing technology defined by DIN 8580. Within the framework of this metalworking process, two or more solid bodies are permanently joined together. It is important that the bodies take on a specific geometric shape through sheet metal processing. Partly this is done by using a "formless material", i.e. the shape of the material is not defined. For example, adhesive counts as a formless substance.

Welding processes for industry and trade.

Joining also has further process groups that belong to metalworking. These are for example soldering, welding, gluing as well as screwing, joining by shearing and upsetting or riveting. In order to establish a connection between the work pieces, further components such as pins, screws, wedges or rivets are sometimes necessary. In addition, specific tools such as soldering irons, impact wrenches and so on are used in this production technology.

Surface Treatment

In metalworking, the term "surface technology" covers all technologies that enable the surface properties of work pieces to be changed. It is important in this respect that the primary function of a work piece in metalworking can be supplemented by surface treatment with additional functions. The principle of functional separation between the surface and the volume of a tool or component is the primary goal of this sheet metal processing manufacturing technique.

The volume of a work piece has a specific function and certain characteristics such as machinability, strength, weight and so on. If a surface technology process is used in sheet metal processing, the work piece or tool can be optimized in such a way that it fulfills a specific requirement profile and has a wide range of functions. These may include the following:

- Barrier function, thermal insulation, diffusion, permeation, corrosion resistance).
- Mechanical protection (friction, wear).
- Optical function (decoration, absorption, reflection).

- Functional integration (functional printing, printing process).
- Electrical function (electrical insulation, conductivity).

Different methods can be used to carry out a surface treatment in sheet metal processing. The surface treatments in metalworking are therefore allocated to different process classes, which are the following:

- Surface removal (electro polishing, burnishing).
- Surface coating (chrome plating, powder coating, painting).
- Mechanical surface treatment (shot peening, polishing, grinding).
- Chemical surface treatment (pickling, degreasing, cleaning).

Machines and Tools for Sheet Metal Working

In the course of sheet metal processing, different tools and machines can be used, whereby a distinction is made between hand tools and power tools. Hand tools include files, iron saws, drawing pins, blind riveting pliers, pliers, hammers, shell irons, and file pistons, ejector pistols and so on. Power tools, on the other hand, include drills, percussion drills, orbital, belt and random orbit grinders, soldering irons, lasers and so on.

Forming

Metal forming is a very important manufacturing operation. It enjoys industrial importance among various production operations due to its advantages such as cost effectiveness, enhanced mechanical properties, flexible operations, higher productivity, considerable material saving.

The objects and articles that we use in our daily life are man-made, engineered parts, which are obtained from some raw material through some manufacturing process. All these objects are made of a number of small components assembled into finished product. The pen that we use for writing, for example is made of several small parts, assembled together. An automobile is supposed to be an assembly of more than 15000 parts, produced through various manufacturing operations.

Manufacturing of finished parts and components from raw materials is one of the most important steps in production. Production encompasses all types of manufacturing processes. Manufacturing refers to the conversion of raw materials into finished products employing suitable techniques. There are several methods of manufacturing such as metal casting, metal forming, metal machining, metal joining and finishing.

Some of the modern methods of manufacturing include micro machining, Nano fabrication, ultra precision manufacturing etc. In order to fulfill the requirements of the ever-increasing demands of various types of industries, the manufacturing engineer has to choose the right type of material and the right type of equipment for manufacture so that the cost of production and the energy consumption are minimum. The selection of suitable manufacturing process should also include concerns for environmental impacts such as air pollution, waste disposal etc.

Modern concepts such as lean manufacturing, adaptive control, agile manufacturing, group technology etc. have considerable influence on cost reduction and quality improvements of products. Computers and robots play important role in modern manufacturing techniques, today. Modeling and simulation of the process prior to mass production helps the manufacturing engineer fix up the best operating parameters and hence achieve the finished product to the utmost level of quality and cost-effectiveness.

Metal Forming

Materials are converted into finished products though different manufacturing processes. Manufacturing processes are classified into shaping [casting], forming, joining, and coating, dividing, machining and modifying material property.

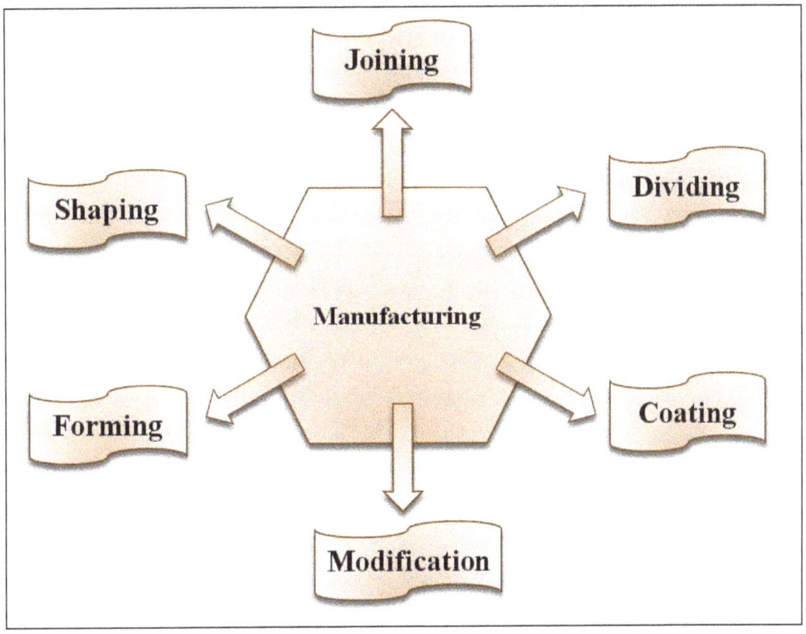

Various manufacturing operations on materials.

Of these manufacturing processes, forming is a widely used process which finds applications in automotive, aerospace, defense and other industries. Wrought forms of materials are produced through bulk or sheet forming operations. Cast products are made through shaping – molding and casting.

A typical automobile uses formed parts such as wheel rims, car body, valves, rolled shapes for chassis, stamped oil pan, etc. In our daily life we use innumerable formed products e.g. cooking vessels, tooth paste containers, bicycle body, chains, tube fitting, fan blades etc.

Forming is the process of obtaining the required shape and size on the raw material by subjecting the material to plastic deformation through the application of tensile force, compressive force, bending or shear force or combinations of these forces.

Classification of Forming

Typically, metal forming processes can be classified into two broad groups. One is bulk forming and the other is sheet metal forming. Bulk deformation refers to the use of raw materials for forming which have low surface area to volume ratio. Rolling, forging, extrusion and drawing are bulk forming processes. In bulk deformation processing methods, the nature of force applied may be compressive, compressive and tensile, shear or a combination of these forces.

Bulk forming is accomplished in forming presses with the help of a set of tool and die. Examples for products produced by bulk forming are: gears, bushed, valves, engine parts such as valves, connecting rods, hydraulic valves, etc.

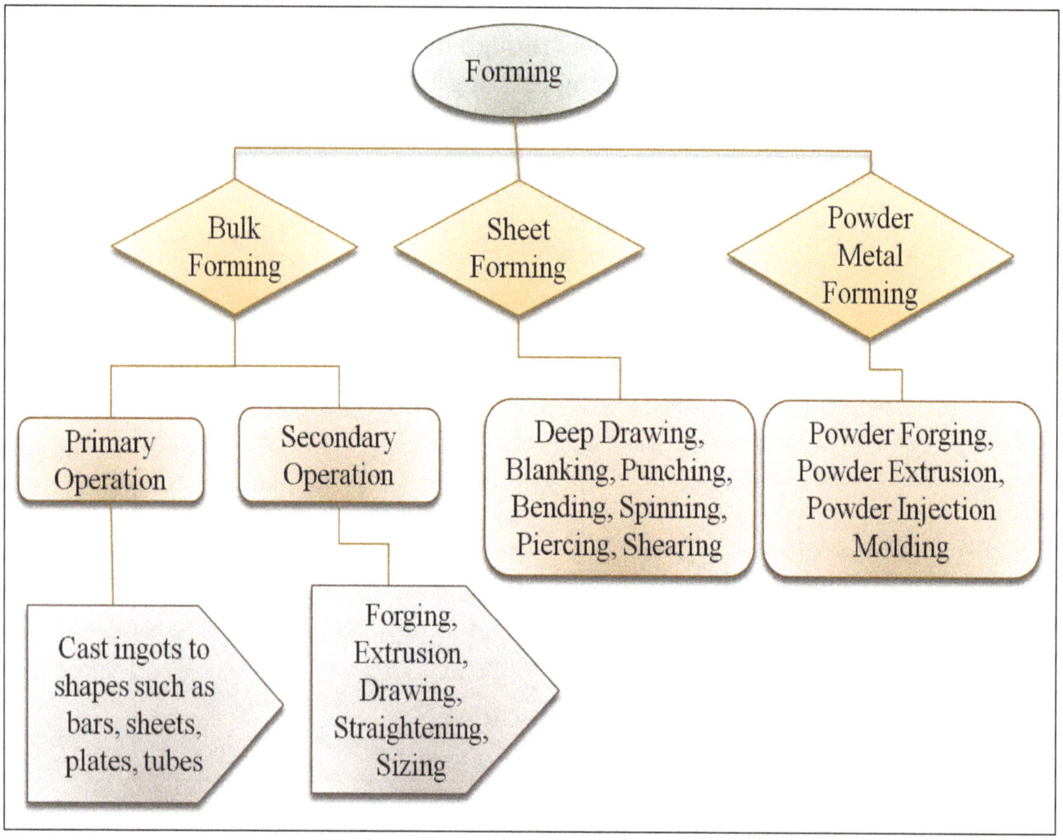

Classification of metal forming processes.

Sheet metal forming involves application of tensile or shears forces predominantly. Working upon sheets, plates and strips mainly constitutes sheet forming. Sheet metal operations are mostly carried out in presses – hydraulic or pneumatic. A set of tools called die and punch are used for the sheet working operations. Bending, drawing, shearing, blanking, punching are some of the sheet metal operations.

A new class of forming process called powder forming is gaining importance due to its unique capabilities. One of the important merits of powder forming is its ability to produce parts very near to final dimensions with minimum material wastage. It is called near-net-shape forming. Material compositions can be adjusted to suit the desirable mechanical properties. Formability of sintered metals is greater than conventional wrought materials. However, the challenge in powder forming continues to be the complete elimination or near-complete elimination of porosity. Porosity reduces the strength, ductility and corrosion resistance and enhances the risk of premature failure of components.

Based on the nature of deformation force applied on the material, during forming, metal forming processes are also classified into several types as shown below:

Forming by compressive stress	Tensile and compressive stresses	Forming under Tensile stress	Bending and shearing stresses
•Open Die Forging •Closed Die Forging •Rolling •Coining •Extrusion	•Deep drawing •Spinning •Stripping •Wrinkle bulging	•Stretch forming •Stretching •Expanding	•Bending •Shearing •Punching •Blanking

Forming is also classified as cold forming, hot forming or warm forming. Hot forming is the deformation carried out at temperatures above recrystallization temperatures. Typically,recrystallization temperatures for materials ranges from 0.5 Tm to 0.8 Tm, where Tm is melting temperature of material.

Forming Operations

Bulk Forming Processes

Rolling is a compressive deformation process, which is used for producing semi-finished products such as bars, sheets, plates and finished products such as angles, channels, sections. Rolling can be carried out both in hot and cold conditions.

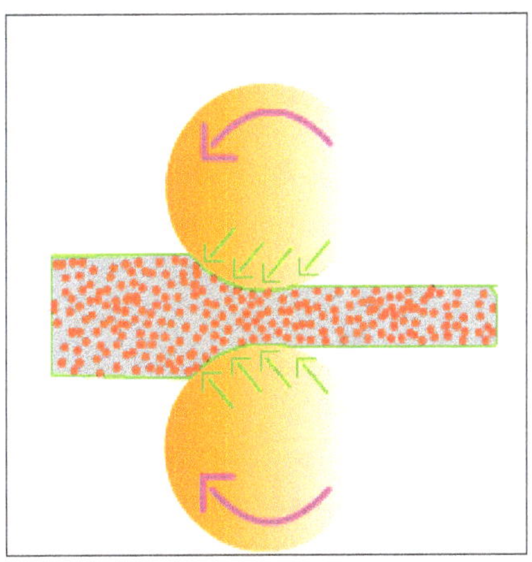

Rolling Process.

Forging is a bulk forming process in which the work piece or billet is shaped into finished part by the application of compressive and tensile forces with the help of a pair of tools called die and punch. Forging can be done in open dies or closed dies. Open die forging is usually used for preliminary shaping of raw materials into a form suitable for subsequent forming or machining.

Open die forming is done using a pair of flat faced dies for operations such as drawing out, thinning, etc.

Closed die forming is performed by squeezing the raw material called billet inside the cavity formed between a pair of shaped dies. Formed products attain the shape of the die cavity.

Valve parts, pump parts, small gears, connecting rods, spanners, etc. are produced by closed die forming.

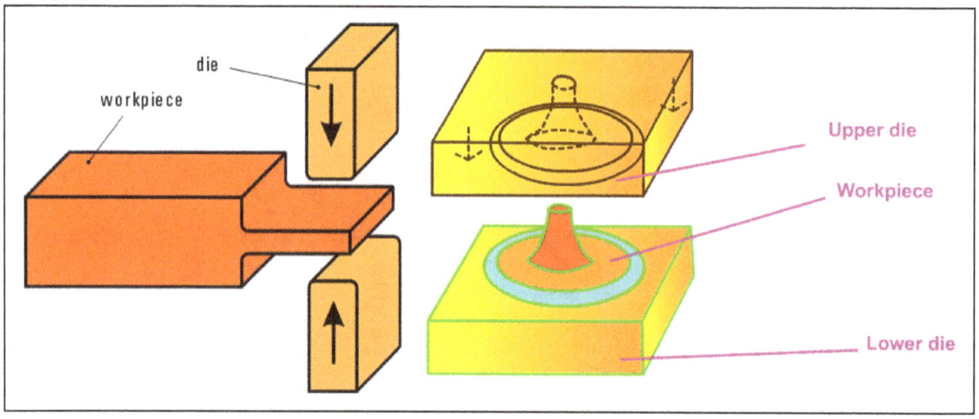

Forging processes.

Coining is the process of applying compressive stress on surface of the raw material in order to impart special shapes on to the surface from the embossing punch – e.g. coins, medallions.

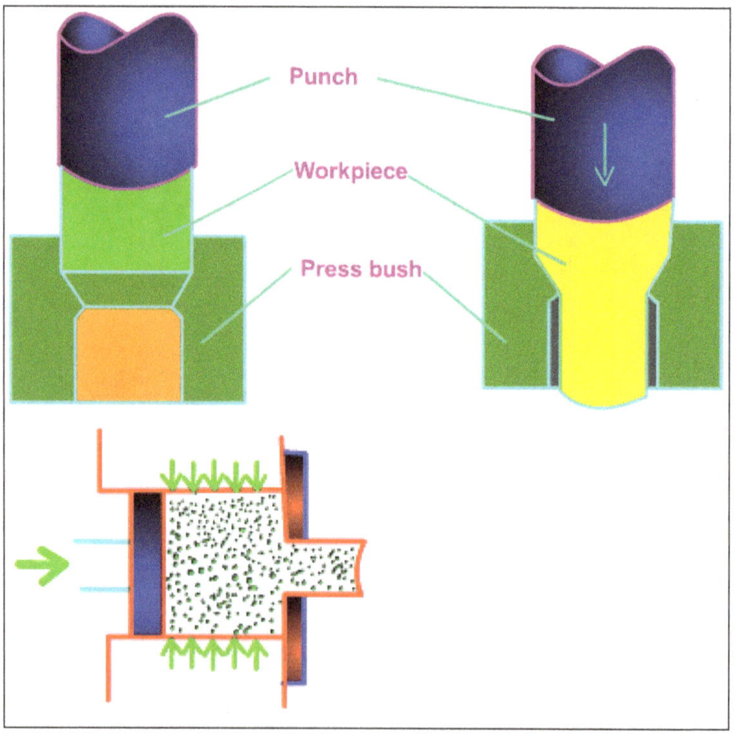

Direct extrusion process.

Extrusion involves forcing the raw material through a narrow opening of constant cross- section or varying cross-section in order to reduce the diameter and increase the length. Extrusion can be done hot or cold. Extruded products include shafts, tubes, cans, cups, gears.

Basically there are two methods of extrusion, forward and backward extrusions. In forward extrusion the work and the extrusion punch move along the same direction. In backward extrusion the punch moves opposite to the direction of movement of the work piece.

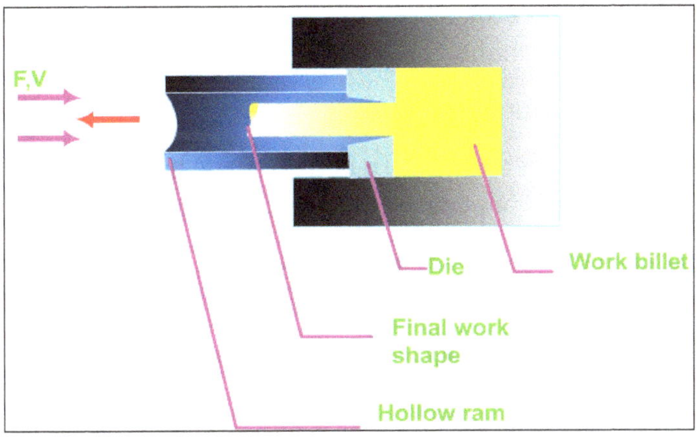

Backward extrusion or Indirect extrusion.

Wire drawing process is used for producing small diameter wires from rods by reducing their diameter and stretching their length through the application of tensile force. Musical strings are produced by wire drawing process. Seamless tubes can be produced by tube drawing process.

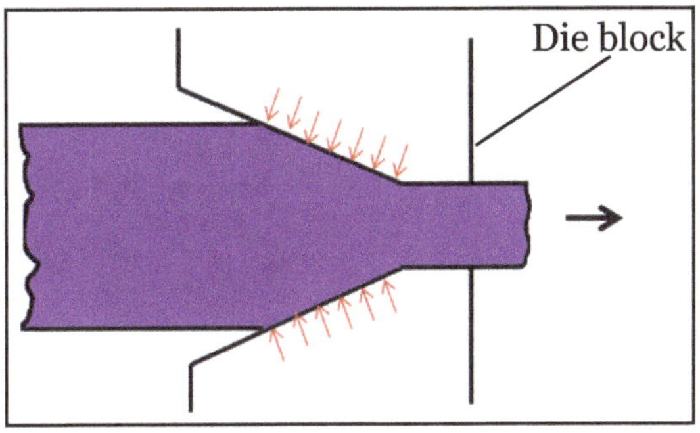

Wire Drawing.

Sheet Metal Operations

Deep drawing is a sheet metal process the process in which a sheet metal is forced into cup of hollow shape without altering its thickness – using tensile and compressive forces. Complex shapes can be produced by deep drawing of blanks in stages – redrawing, multiple draw deep drawing etc.

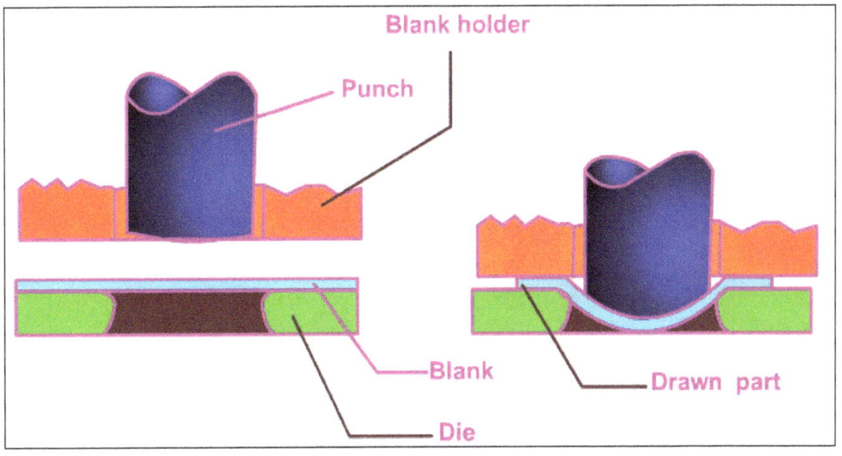

Deep drawing.

- Hydro mechanical deep drawing uses both punch force and hydrostatic force of a pressurized fluid for achieving the shape.
- Flanges and collars are formed by flanging process.
- Spinning transforms a sheet metal into a hollow shape by compressive and tensile stresses. Spinning mandrel of given shape is used against a roll head.
- Embossing imparts an impression on the work piece by means of an embossing punch. Bending of sheets includes rotary bending, swivel bending, and roll bending using rotary die.
- Die bending using flat die or shaped die is used for bending of sheets, or die coining of sheets.

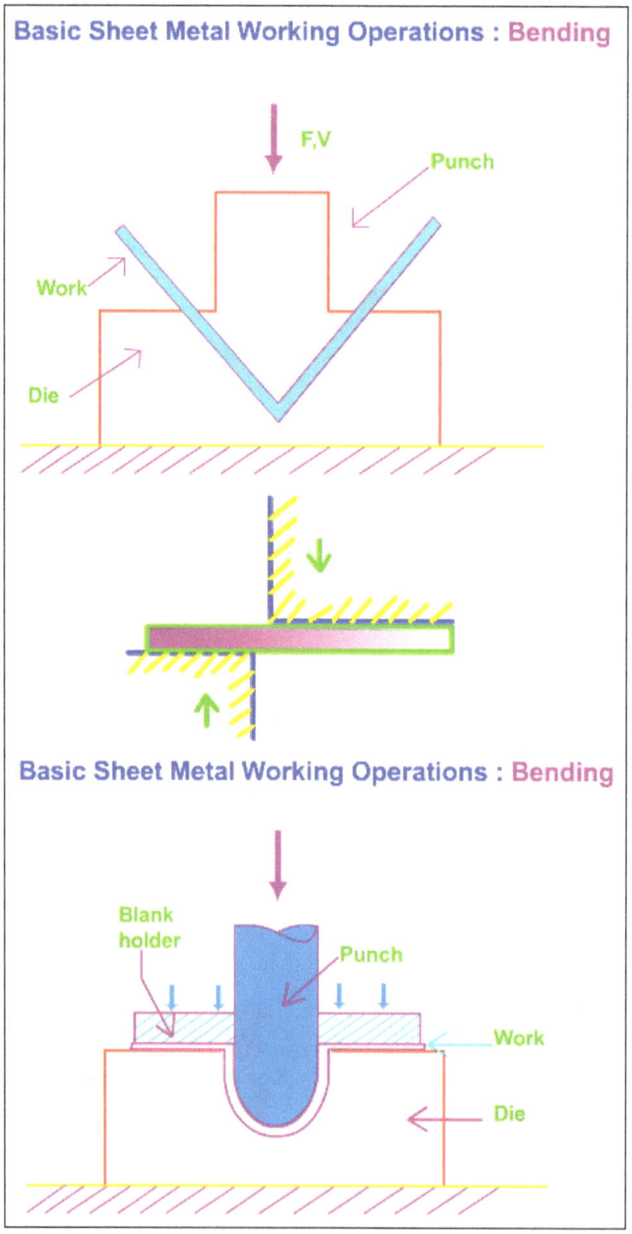

Bending and shearing.

Cutting

Metal cutting is "the process of removing unwanted material in the form of chips, from a block of metal, using cutting tool". A person who specializes in machining is called a machinist. A room, building or company where machining is done is called a Machine Shop.

The basic elements involved in this process are:

- A block of metal (work piece).
- Cutting Tool.
- Machine Tool.
- Cutting Fluid.
- Cutting speed (Primary Motion).
- Feed (Secondary Motion).
- Chips.
- Work holding and Fixturing.
- Force and Energy Dissipated, and
- Surface Finish.

The essential conditions for successful metal cutting are:

- Relative motion between work and cutting tool.
- Tool material must be harder than work material.
- Work and tool must be rigidly held by jig and fixtures.
- Sharp Cutting edge of cutting tool.
- Primary Motion (Cutting Speed).

Almost all the products produced by metal removal process, either directly or indirectly. The major disadvantages of the process are loss of material in the form of chips.

Forging

Forging is a manufacturing process involving the shaping of a metal through hammering, pressing, or rolling. These compressive forces are delivered with a hammer or die. Forging is often categorized according to the temperature at which it is performed—cold, warm, or hot forging.

A wide range of metals can be forged. Typical metals used in forging include carbon steel, alloy steel, and stainless steel. Very soft metals such as aluminum, brass, and copper can also be forged. The forging process can produce parts with superb mechanical properties with minimum waste. The basic concept is that the original metal is plastically deformed to the desired geometric shape—giving it higher fatigue resistance and strength. The process is economically sound with the ability to mass produce parts, and achieve specific mechanical properties in the finished product.

Forging involves the shaping of metal through compressive forces such as hammering, pressing, or rolling.

History of Forging

Forging has been practiced by smiths for thousands of years. At first, bronze and copper were the most common forged metals, in the Bronze Age: later, as the ability to control temperature and the process of smelting iron was discovered, iron became the primary forged metal. Traditional products include kitchenware, hardware, hand tools, and edged weapons. The Industrial Revolution allowed forging to become a more efficient, mass-production process. Since then, forging has evolved along with advances in equipment, robotics, electronic controls, and automation. Forging is now a worldwide industry with modern forging facilities producing high-quality metal parts in a vast array of sizes, shapes, materials, and finishes.

Metal is heated before it is manipulated to the desired shape using a forging hammer. This used to be done by hand by blacksmiths.

Forging Methods

There are several forging methods with different capabilities and benefits. The more commonly used forging methods include the drop forging methods, as well as roll forging.

Drop Forging

Drop forging derives its name from the process of dropping a hammer onto the metal to mold it into the shape of the die. The die refers to the surfaces that come into contact with the metal. There are two types of drop forging—open-die and closed-die forging. Dies are typically flat in shape with some having distinctively shaped surfaces for specialized operations.

Open-Die Forging (Smith Forging)

Open-die forging is also known as smith forging. A hammer strikes and deforms a metal on a stationary anvil. In this type of forging, the metal is never completely confined in the dies—allowing it to flow except for the areas where it is in contact with the dies. It is the operator's responsibility to orient and position the metal to achieve the desired final shape. Flat dies are used, with some having specially shaped surfaces for specialized operations. Open-die forging is suitable for simple and large parts, as well as customized metal components.

Advantages of open-die forging:

- Better fatigue resistance and strength.
- Reduces chance of error and/or holes.
- Improves microstructure.
- Continuous grain flow.
- Finer grain size.

Closed-Die Forging (Impression-Die)

Closed-die forging is also known as impression-die forging. The metal is placed in a die and attached to an anvil. The hammer is dropped onto the metal, causing it to flow and fill the die cavities. The hammer is timed to come into contact with the metal in quick succession on a scale of milliseconds. Excess metal is pushed out from the die cavities, resulting in flash. The flash cools faster than the rest of the material, making it stronger than the metal in the die. After forging, the flash is removed.

In order for the metal to reach the final stage, it is moved through a series of cavities in a die:

- Edging impression (also known as fullering or bending): The first impression used to mold the metal into a rough shape.
- Blocking cavities: The metal is worked into a shape that more closely resembles the final product. The metal is shaped with generous bends and fillets.
- Final impression cavity: Final stage of finishing and detailing the metal into the desired shape.

Advantages of closed-die forging:

- Produces parts up to 25 tons.
- Produces near net shapes that require only a small amount of finishing.
- Economic for heavy production.

Roll Forging

Roll forging consists of two cylindrical or semi-cylindrical horizontal rolls that deform a round or flat bar stock. This works to reduce its thickness and increase its length. This heated bar is inserted and passed between the two rolls—each containing one or more shaped grooves—and is progressively shaped as it is rolled through the machine. This process continues until the desired shape and size is achieved.

Advantages of automatic roll forging:

- Produces little to no material waste.
- Creates a favorable grain structure in the metal.
- Reduces the cross-sectional area of the metal.
- Produces taper ends.

Press Forging

Press forging uses a slow, continuous pressure or force, instead of the impact used in drop-hammer forging. The slower ram travel means that the deformation reaches deeper, so that the entire volume of the metal is uniformly affected. Contrastingly, in drop-hammer forging, the deformation is often only at the surface level while the metal's interior stays somewhat undeformed. By controlling the compression rate in press forging, the internal strain can also be controlled.

Press forging uses a slow, continuous pressure or force to shape metal uniformly, instead of the impact used in drop forging.

Advantages of press forging:

- Economic for heavy production.

- Greater accuracy in tolerances within 0.01–0.02 inch.
- Dies have less draft allowing for better dimensional accuracy.
- Speed, pressure, and travel of the die are automatically controlled.
- Process automation is possible.
- Capacity of presses range from 500–9000 tons.

Upset Forging

Upset forging is a manufacturing process that increases the diameter of the metal by compressing its length. Crank presses, a special high-speed machine, are used in upset forging processes. Crank presses are typically set on a horizontal plane to improve efficiency and the quick exchange of metal from one station to the next. Vertical crank presses or a hydraulic press are also options.

Advantages of upset forging:

- High production rate of up to 4500 parts per hour.
- Full automation is possible.
- Elimination of the forging draft and flash.
- Produces little to no waste.

Automatic Hot Forging

In automatic hot forging, mill-length steel bars are inserted into one end of the forging machine at room temperature, and hot forged products emerge from the other end. The bar is heated with high-power induction coils to a temperature ranging from 2190–2370°F in under 60 seconds. The bar is descaled with rollers and shared into blanks. At this point, the metal is transferred through several forming stages that can be coupled with high-speed cold-forming operations. Typically, the cold-forming operation is left for the finishing stage. By doing so, the benefits of cold-working can be reaped while also maintaining the high speed of automatic hot forging.

Advantages of automatic hot forging:

- High output rate.
- Acceptance of low-cost materials.
- Minimal labor required to operate machinery.
- Produces little to no material waste (material savings between 20–30% over conventional forging).

Precision Forging (Net-Shape or Near-Net-Shape Forging)

Precision forging requires little to no final machining. It is a forging method developed to minimize the cost and waste associated to post-forging operations. Cost savings are achieved from the reduction of material and energy, as well as the reduction of machining.

Isothermal Forging

Isothermal forging is a forging process where the metal and die are heated to the same temperature. Adiabatic heating is used–there is no net transfer of mass or thermal exchange between the system and the external environment. The changes are all due to internal changes resulting in highly controlled strain rates. Due to the lower heat loss, smaller machines may be used for this forging process.

Metalworking Machinery

Metalworking machinery accessories are devices that are used as parts of machine tools and are important to the machining of metals. These can be broadly classified into work holding, metalworking fluids, and power tools accessories. The global metalworking machinery accessories market is segmented based on products and geography.

Such machines are heavy, non-portable, power-driven equipment enabling manufacturers to mass produce numerous metal components or finished products. Such components produced using metalworking machines can be assembled into heavy machinery, automobiles and other machinery equipment. Metalworking machines are categorized into two categories namely, metal forming and metal cutting. Metalworking machines comprise manufacturing machinery and machine tools which are used to cut or shape metals and other hard materials to produce industrial tools.

Metal forming machines are used to squeeze metals into specific shapes by applying processes including cold-heading, bending, coil slitting, shearing, pressing and stamping. Such metal forming machines enable manufacturers to design metal components as per specific requirements which can used for specific tasks or incorporate into larger machinery. Metal cutting machines are used to cut away swarf or chips of metals to shape metal work piece into desired product. Metal cutting machines comprise drilling machines, broaching machines, gear cutting machines, electrical-discharge machines and machining centers. Other metal cutting machines include grinders, transfer machines, milling machines and turning machines.

Metalworking machines comprise manufacturing machinery and machine tools which are used to cut or shape metals and other hard materials to produce industrial tools. Metalworking machines use a combination of techniques including forming, cutting and physio-chemical processing to design and manufacture specific components or goods.

On account of its immense capabilities, metalworking machines market has become lucrative in recent years. With increasing demand for machining centers and machining tools, numerous players in the metalworking machines market have invested heavily to design and develop such machinery. Moreover, there has been an increasing demand for customized machinery equipment which is further driving the growth of metalworking machines market. Companies in this market conduct research and development activities to identify and cater to various unique manufacturing processes.

There are lots of machines and tools that make a well addition to the metalworking shop and unlock new methods of metalworking possibilities of metal treatment. In applications of electronics, robotics, metal or woodworking, machining, welding, metals mithing, sewing, or any other number of actions, these machines are very helpful to complete tasks which include workings by Bandsaw, hacksaw, lathe machine, drilling etc.

Metalworking industries employ a wide range of machine tools to create parts that are of a desired shape and size. These machines have evolved over the past hundreds of years and have made it possible to produce not only the most complex shapes but also in a way that involves very less human effort.

There are various techniques used by the fabrication industry that has led to invention and evolution of many machines. Following are the most widely used machines in the metalworking industry:

- Lathe Machine: These are the machines that rotate the workpiece to perform various machining operations. Lathes can be used to produce parts that are ferrous, non-ferrous metals, plastics, and wood. They are multipurpose machines that can be used to perform operations like cutting, sanding, knurling, drilling, boring, threading, facing and turning. They come in various sizes, and with a range of accessories.
- Milling Machine: It is a machine that uses rotary cutters to remove the undesired material. The workpiece is held tight on the movable worktable and a rotating tool is made to move vertically. The two major operations that a milling machine performs is face milling and peripheral milling.
- Grinding Machine: In this machine, an abrasive wheel is made to rotate at relatively high speeds. The rotating wheel provides finishing to the workpiece by removing a very small amount of material along its surface, after which it becomes smooth to touch.

- Drilling Machines: These are one of the most common types of machines that we can see around us. They are used for various household repair works too. The tools of this machine are easy to install and replace which makes it really easy to use. They can be used for making a hole in a wide range of materials which not only includes metals but also materials like cement and concrete.
- Shaper Machines: It uses a single point cutting tool that moves in a linear motion to perform a cutting operation on a workpiece. It has a simple mechanism and is easy to operate too.
- Broaching Machine: It uses a toothed tool or a broach to remove material from the workpiece. There are two types of broaching processes – linear and rotary.
- Saw machines: There are three types of saw machines, hacksaw, circular saw and band saw. These machines are used to cut a metal piece into two. It uses a blade which is either rotating or reciprocating to make the cut.
- Planer Machine: This machine is very similar to shaper machine and the only difference in this machine is, the cutting tool is stationary and the workpiece is made to move in a linear motion. Generally, these machines are larger than shaper machines.
- Shearing Machine: It is a machine that is used for cutting sheet metal. Shearing process does not form chips or any residual material. Other major metal cutting operations that are similar to this are blanking and piercing.
- Hobbing Machines: These machines are used for making gears and splines. It is a type of milling machine that uses a special type of cutting tool but is relatively inexpensive and is mostly used to produce spur gears.
- Drill Press: It is a more reliable version of a hand-held drill as they have better stability than them. The workbench allows the drills to be made continuously with good accuracy.

Metalworking Fluids

Metalworking fluid (MWF) is the name given to a range of oils and other liquids that are used to cool and/or lubricate metal work pieces when they are being machined, ground, milled, etc. MWFs reduce the heat and friction between the cutting tool and the workpiece, and help prevent burning and smoking. Applying MWFs also helps improve the quality of the workpiece by continuously removing the fines, chips, and swarfs from the tool being used and the surface of the workpiece. (Swarfs are the small pieces of metal removed from a workpiece by a cutting tool.).

Different Types of MWFs

While there are many different components and additives in MWFs, there are four basic classes:

- Straight Oils: Also called "cutting" or "neat" oils. This type is made up of mineral (petroleum), animal, marine, vegetable or synthetic oils. Today, the mineral oils are "severely solvent refined" or "severely hydrotreated". These terms refer to refining processes that help reduce the amount of polynuclear aromatic hydrocarbons (PAHs). Straight oils are not diluted with water but other additives may be present.

- Soluble Oils (emulsifiable oils): This category contains 30 to 85 percent severely refined petroleum oils, as well as emulsifiers to disperse the oil in water.
- Semi-synthetic fluids: This category contains 5 to 30 percent severely refined petroleum oils, 30 to 50 percent water and a number of additives.
- Synthetic fluids: This category does not contain petroleum oils. Instead, they use detergent-like components and other additives to help "wet" the workpiece.

Although each class will vary greatly in composition, each may contain additives such as:

- Sulphurized or chlorinated compounds.
- Corrosion inhibitors (e.g., calcium sulfonate, sodium sulfonates, fatty acid soaps, amines, boric acid).
- Extreme pressure additives (e.g., sulfurized fatty materials, chlorinated paraffins, phosphorus derivatives).
- Anti-mist agents (e.g., polyisobutylene polymer).
- Anti-weld agents.
- Emulsifiers (e.g., triethanolamine, sodium petroleum sulphonates, salts of fatty acids and non-ionic surfactants).
- Alkanolamines.
- Biocides (e.g., triazine compounds, oxazolidine compounds).
- Preservatives.
- Stabilizers.
- Dispersants.
- Defoamer.
- Colourants.
- Dyes.
- Odourants.
- Fragrances.

Does the Composition of MWFs Change with Storage or use?

When stored, nitrosamines can form while the fluid is stored for long periods of time. Nitrosamines form slowly in the water-based MWFs and may be the result of interaction of nitrites in the fluid, lining of the cans used for storage, or from nitrogen oxides in air. Recycling MWFs can increase the problem if more reactants are added. The formation of nitrosamines in the metalworking fluids is a concern since many nitrosamines are classified as carcinogens.

When MWFs are used, a primary concern is the presence of contaminants that encourage the growth of bacteria and fungi in water-based MWFs. The bacteria can degrade the emulsions and change the properties of the MWFs. While biocides are added to reduce the amount of microbial growth, the biocide products themselves have hazardous properties.

Other sources of contamination include "tramp" oil - oil used for lubrication of the machines, such as hydraulic oil, gear box oil, and other lubricants. Tramp oils that leak into the metalworking fluids can contribute to microbial growth in many ways, including by being a source of nutrients for bacteria, and by creating various conditions for anaerobic microbial growth.

MWFs are also contaminated by small particles of the metal or alloy objects (e.g., fines, chips, swarfs) that come off the parts while they are being machined. Common metals used include steel or alloys containing nickel, cobalt and chromium.

In addition, while the extent of the problem is not clear, there is the potential for straight oils to be heated during use (usually at the site where the cutting tool works on the metal workpiece), and the temperature may increase high enough to cause the formation of polynuclear hydrocarbons (or polyaromatic hydrocarbons, PAHs).

MWFs may also be contaminated by water, cleaning products used for routine housekeeping, or other products at the work site. Improper recycling of materials or the addition of unspecified fluids (such as old lubricating oils) to the MWF will also change the composition of fluid.

How do MWFs Enter the Body?

MWFs can enter the body when:

- The mist, aerosols, or vapor is inhaled. Exposure will depend on:
    - What kind of machining being done.
    - How the fluid is applied (e.g., manually with an oil can; flooded through a hose or pipe, or atomized (aerosolized) and the mist directed where the tool contacts the workpiece).
    - How, or if, the machine is enclosed and ventilated. Higher exposures happen when:
        - The operator works close to the metalworking machine.
        - The operations involve high-speed tools or deep cuts.
        - The machines do not have an enclosed process.
        - The ventilation is poor.
- It comes into contact with the skin. The risk of absorption through the skin is high especially if there are cuts, rashes, cracks, or other breaks in the skin. Hands and arms are most at risk if adequate precautions are not taken. Fluids can splash onto the skin during

machining, and can also occur when they are prepared or drained, when work pieces are handled, when tools are being changed and set, and during maintenance or cleaning operations. Rags or clothes soaked with MWF that are in constant contact with the skin are also a concern (including when rags are placed in the pockets of clothing).

- Ingested if you eat, drink or smoke at the workstation or without washing your hands first.

Health Concerns About Working with MWFs

MWFs have been associated with several health concerns. Contact dermatitis is the most common skin condition reported. Symptoms include burning, itching, and blistery skin. Inhalation of the mists may cause asthma and lung irritation (hypersensitivity pneumonitis), chronic bronchitis, and impaired lung function.

There is also evidence that some MWFs are associated with an increased risk of certain cancers such as larynx, rectum, pancreas, skin, scrotum, and bladder. Since the time between exposure and the development of the disease is often more than 20 years, most of the cancer cases associated with the use of MWFs is due to the exposure to MWFs used in the mid-1970s or earlier. In the past few decades, substantial changes to the composition of MWFs and reductions in the contaminants have occurred. As a result, the risk of cancer from more recent exposures is not as clear.

Skin

All types of MWFs can cause skin irritation. If you had severe eczema as a child, there is a high risk that you will suffer dermatitis when exposed to MWFs. Exposure occurs when hands are dipped into the fluid, or when a person handles the parts, tools and equipment covered in fluid. Splashing is a concern if guarding is absent or inadequate. Clothing contaminated with MWF, poor housekeeping, and poor personal hygiene also contributes to skin exposure.

Irritant or allergic contact dermatitis is reported to occur from exposure to soluble, semisynthetic, and synthetic MWFs. Dermatitis can be caused by:

- Bacteria and their by-products.
- Chemicals added to control bacteria (biocides).
- Chemicals added to control rust and corrosion.
- Contact with metal contaminants such as nickel, cobalt and chromium, which are known sensitising agents.

Skin conditions associated with straight oils include:

- Folliculitis (inflammation of hair roots or follicles). Caused by prolonged and regular contact with straight oils.
- Oil acne (red bumps with yellow pustules/ blisters filled with pus). Caused by skin contact with oil soaked clothing. Can develop on various body parts: face, forearms, thighs, legs, etc.
- Irritation. The small metal particles (fines and swarfs), generated while parts are machined, can damage the skin and make existing irritations worse.

In addition, the small metal particles (fines and swarfs) generated while parts are machined can damage the skin and make existing irritations worse. Skin conditions can become disabling if not treated or if the worker continues to work with condition.

Respiratory

An increase in the number of work-related asthma, bronchitis, irritation of the respiratory tract and breathing difficulties has been reported among those exposed to MWFs. Exposure to mist, aerosol, and vapour can lead to the development of respiratory conditions or can aggravate the existing ones.

It is not clear whether respiratory problems are caused by specific fluid components, contaminants, products of microbial growth or degradation, or a combination of these factors. For example, metalworking fluid-induced asthma is reported more consistently with synthetic MWFs, but MWF-induced asthma can also occur with soluble and straight fluids.

Again, the severity of exposure depends on proximity to the machine, and if the operations involve high tool speeds and deep cuts, if the machine is enclosed, or if ventilation equipment is working properly. High pressure or excessive fluid application, contamination of fluid (with tramp oil), improper fluid selection, and poor maintenance will also result in higher exposures.

Bacteria contamination can cause irritation of respiratory tracts or flu-like symptoms, aggravate asthma, and irritate the eyes, nose and throat (causing a sore throat, red watery itchy eyes, runny nose, nose bleeds, coughing wheezing, or shortness of breath). For example, hypersensitivity pneumonitis (HP) is an allergic type reaction in the lungs that may be caused by exposure to microbial products. HP is marked by chills, fever, shortness of breath and a deep cough - similar to a cold that will not go away. If left untreated, it can lead to irreversible lung damage.

Cancer

Cancers often associated with exposure to metalworking fluids include rectum, pancreas, larynx, skin, scrotum, esophagus, and bladder. The National Institute for Occupational Safety and Health (NIOSH) in the USA reports that studies were not highly consistent regarding the specific types of cancer associated with MWFs. This uncertainty is likely due to the wide variation in the types of MWFs and contaminants and the lack of detailed exposure information.

Also, because the latency period (the time between first exposure and the discovery of disease) for cancer is often 20 years or more, it is likely that the diseases studied recently are associated with older formulations of MWFs (from the mid-1970s and earlier). For example, fluids used before 1985 may have contained nitrites, mildly refined petroleum oils and other chemicals which were removed because of health concerns. Cancer risks have likely been reduced, but there is not enough data yet to prove this theory.

Areas of concern for risk of cancer currently include:

- Unrefined mineral oils and contact with exposed skin (including oil soaked clothing and especially oily rags kept in pockets, which caused cancer of scrotum).

- Nitrites or nitrates and amines that cause the formation of nitrosamines when MWFs are heated or under pressure. Certain nitrosamines, such as N-nitrosodiethanolamines (NDELA), are known to be cancer causing agents.
- Some biocides release formaldehyde, a suspected carcinogen. Formaldehyde can also speed up the formation of nitrosamines.
- Chlorinated paraffins are carcinogens (often used when extreme pressure is required). They also form dioxin, another carcinogen.

How do you find out about the Composition of a MWF?

The supplier/manufacturers of the fluid can provide you with the Safety Data Sheet (SDS) which will provide information about the ingredients and health and safety hazards. It is important to monitor the MWF for contamination, and to have good work practices that help keep the fluids as free from contamination as possible.

How can you work Safely with MWFs?

MWFs, in general, may contain any number and concentration of hazardous components. The risk of exposure to these chemicals varies with the manufacturing process, as well as changes such as refining, recycling, degradation, or using reclaimed chemicals, and potential reactions between components.

Since there are so many varieties of MWFs, it is best to work with them safely and keep exposures as low as possible, no matter which type is being used. Steps include:

- Obtain Safety Data Sheets (SDSs) from the supplier so you know exactly what type of MWF you are working with and what precautions to take.
- Obtain technical bulletins that may provide additional health and safety information.
- Apply exposure control measures.

Exposure Control Measures

Substitution

- Choose MWFs with the least toxic materials, whenever possible.

Engineering Controls

Design and Operation

Fine mists are created when the MWF stream breaks up during use and becomes airborne, especially when the fluid is moved at a high speed or velocity. The small mist droplets are easily suspended in air and are hard to contain or collect. To minimize the amount of mist produced it is important to choose fluid delivery systems which release minimum amount of mist.

Ways to reduce the amount of mist include the following:

- Low pressure delivery of MWF.

- Addition of mist suppressants.
- Lower MWF flow rate.
- Covered fluid reservoirs and return system to prevent contamination.
- Proper machine maintenance (e.g., no leaks causing contamination).
- Interrupt (stop) the flow of MWF when a part is not being machined instead of the fluid running continuously (e.g., feeding the machine with MWF only when parts are being machined).
- Do not use compressed air to blow clean parts covered in MWF as the air pressure will cause the fluid to become airborne.

Effective Ventilation

- Exhaust ventilation prevents accumulation and recirculation of contaminants.
- Local exhaust (near the source) is the most effective.
- Enclosed operations are easier to ventilate.

Isolation

- Install complete enclosures or splash guards, depending on the operation, to keep the metalworking fluids contained and away from the operator.

Proper use of Biocides

- Use biocides according to supplier or manufacturer's directions. Overuse of biocides can cause biocide-resistant strains to develop or another strain to overtake other strains.
- Biocides themselves can cause either allergic or contact dermatitis.

Administrative Controls

Good work practices include the following:

Proper Maintenance of Equipment

- Reduce the amount of contaminants into the MWFs, such as hydraulic oils and other "tramp" oils, by keeping equipment in good working order.
- Make sure that all systems (ventilation, guarding, etc.) are maintained properly.

Appropriate Level of Personal Hygiene

- Stress the importance of personal hygiene. To maintain clean skin, be sure to wash with gentle soaps, use clean water and towels and wear clean work clothes (those that are not soaked in fluids).
- To prevent unintentional ingestion, do not eat, drink or smoke in the work area, and always wash your hands before eating, drinking or smoking. Observe good hygiene - wash your hands before and after you go to the bathroom.

- Barrier creams developed for specific hazards may offer a level of protection, but they should not be substituted for good personal hygiene and chemical protective gloves. The effectiveness of barrier creams has not been well documented. Some barrier creams can actually make some skin conditions worse so they should only be applied to normal, healthy skin (that is, no cuts, rashes, scratches, etc.).

Good Housekeeping

- Maintain good housekeeping. Keep floors, equipment and the general work environment clean. Use appropriate cleaning agents, work practices, and protective clothing. All workers should be trained in how to clean MWFs properly.
- Spills should be cleaned immediately. Wastes, including floor wash water, should not be dumped or swept into the MWF sumps or coolant return trenches. Solvent soaked rags should be deposited in airtight metal containers.
- All machines should be cleaned and MWF changed periodically. When changing the MWF, thoroughly clean the entire system to remove bacterial deposits.

Personal Protective Equipment (PPE)

Engineering controls are preferred before using PPE, but in certain situations, PPE may be required. Employees should be trained to know when PPE are necessary, what PPE to wear, how to wear and remove it properly, the limitations of the PPE, and its proper care and maintenance.

PPE that may be required when working with metalworking machines and metalworking fluids include those that provide protection from:

- Chemicals in the MWFs, cleaning fluids, etc.
- Flying metal particles (fines and swarfs).
- Sharp edged parts.
- High temperatures / hot parts that could produce burns.
- Falling objects.
- Machine noise.

For example, gloves, protective sleeves, aprons, eye protection (goggles and/or face shields with safety goggles), chemical-resistant clothing, and caps may be needed. However, in some situations, gloves may not be appropriate as they can get tangled in moving parts or work pieces. A thorough hazard assessment of the task must be done.

Respiratory protection that is classified as "resistant to oil" (class R) or oil proof (class P) should be selected where appropriate. Depending on the level of airborne contaminants, an air-purifying, half mask respirator (with HEPA filter) including disposable (P- or R-series) (for oil mists less than 50 mg/m^3), or any powered, air-purifying respirator equipped with hood or helmet and HEPA filter (for oil mists less than 125 mg/m^3).

3

Metallurgical Concepts

Concepts like stoichiometry, thermochemistry, material balance, mineral beneficiation, etc. all fall under the umbrella of metallurgy. This chapter addresses such metallurgical terms and discusses them in detail for an extensive understanding of the subject.

Stoichiometry

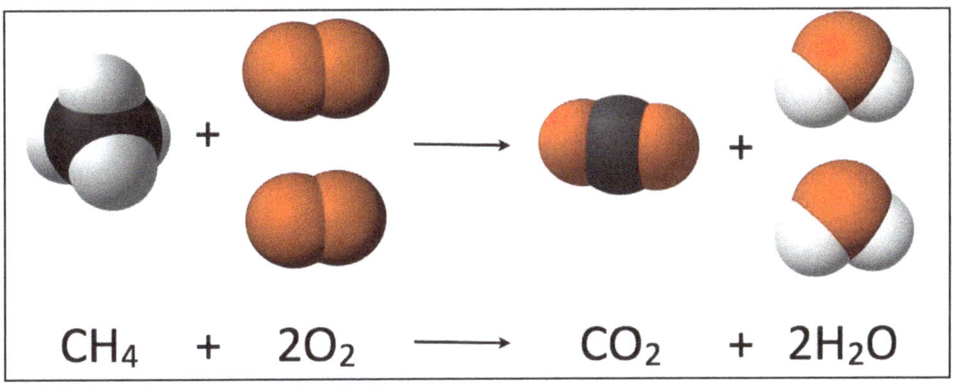

Stoichiometry is the calculation of relative quantities of reactants and products in chemical reactions.

Stoichiometry is founded on the law of conservation of mass where the total mass of the reactants equals the total mass of the products leading to the insight that the relations among quantities of reactants and products typically form a ratio of positive integers. This means that if the amounts of the separate reactants are known, then the amount of the product can be calculated. Conversely, if one reactant has a known quantity and the quantity of product can be empirically determined, then the amount of the other reactants can also be calculated.

This is illustrated in the image here, where the balanced equation is:

$$CH_4 + 2\ O_2 \quad CO_2 + 2\ H_2O.$$

Here, one molecule of methane reacts with two molecules of oxygen gas to yield one molecule of carbon dioxide and two molecules of water. Stoichiometry measures these quantitative relationships, and is used to determine the amount of products/reactants that are produced/needed in a given reaction. Describing the quantitative relationships among substances as they participate in chemical reactions is known as *reaction stoichiometry*. In the example above, reaction stoichiometry measures the relationship between the methane and oxygen as they react to form carbon dioxide and water.

Because of the well known relationship of moles to atomic weights, the ratios that are arrived at by stoichiometry can be used to determine quantities by weight in a reaction described by a balanced equation. This is called *composition stoichiometry*.

Gas stoichiometry deals with reactions involving gases, where the gases are at a known temperature, pressure, and volume and can be assumed to be ideal gases. For gases, the volume ratio is ideally the same by the ideal gas law, but the mass ratio of a single reaction has to be calculated from the molecular masses of the reactants and products. In practice, due to the existence of isotopes, molar masses are used instead when calculating the mass ratio.

Definition

A stoichiometric amount or stoichiometric ratio of a reagent is the optimum amount or ratio where, assuming that the reaction proceeds to completion:

1. All of the reagent is consumed.
2. There is no deficiency of the reagent.
3. There is no excess of the reagent.

Stoichiometry rests upon the very basic laws that help to understand it better, i.e., law of conservation of mass, the law of definite proportions (i.e., the law of constant composition), the law of multiple proportions and the law of reciprocal proportions. In general, chemical reactions combine in definite ratios of chemicals. Since chemical reactions can neither create nor destroy matter, nor transmute one element into another, the amount of each element must be the same throughout the overall reaction. For example, the number of atoms of a given element X on the reactant side must equal the number of atoms of that element on the product side, whether or not all of those atoms are actually involved in a reaction.

Chemical reactions, as macroscopic unit operations, consist of simply a very large number of elementary reactions, where a single molecule reacts with another molecule. As the reacting molecules (or moieties) consist of a definite set of atoms in an integer ratio, the ratio between reactants in a complete reaction is also in integer ratio. A reaction may consume more than one molecule, and the stoichiometric number counts this number, defined as positive for products (added) and negative for reactants (removed).

Different elements have a different atomic mass, and as collections of single atoms, molecules have a definite molar mass, measured with the unit mole (6.02×10^{23} individual molecules, Avogadro's constant). By definition, carbon-12 has a molar mass of 12 g/mol. Thus, to calculate the stoichiometry by mass, the number of molecules required for each reactant is expressed in moles and multiplied by the molar mass of each to give the mass of each reactant per mole of reaction. The mass ratios can be calculated by dividing each by the total in the whole reaction.

Elements in their natural state are mixtures of isotopes of differing mass, thus atomic masses and thus molar masses are not exactly integers. For instance, instead of an exact 14:3 proportion, 17.04 kg of ammonia consists of 14.01 kg of nitrogen and 3 × 1.01 kg of hydrogen, because natural nitrogen includes a small amount of nitrogen-15, and natural hydrogen includes hydrogen-2 (deuterium).

A stoichiometric reactant is a reactant that is consumed in a reaction, as opposed to a catalytic reactant, which is not consumed in the overall reaction because it reacts in one step and is regenerated in another step.

Converting Grams to Moles

Stoichiometry is not only used to balance chemical equations but also used in conversions, i.e., converting from grams to moles using molar mass as the conversion factor, or from grams to milliliters using density. For example, to find the amount of NaCl (sodium chloride) in 2.00 g, one would do the following:

$$\frac{2.00\ \text{g NaCl}}{58.44\ \text{g NaCl mol}^{-1}} = 0.034\ \text{mol}$$

In the above example, when written out in fraction form, the units of grams form a multiplicative identity, which is equivalent to one (g/g = 1), with the resulting amount in moles (the unit that was needed), as shown in the following equation,

$$\left(\frac{2.00\ \text{g NaCl}}{1}\right)\left(\frac{1\ \text{mol NaCl}}{58.44\ \text{g NaCl}}\right) = 0.034\ \text{mol}$$

Molar Proportion

Stoichiometry is often used to balance chemical equations (reaction stoichiometry). For example, the two diatomic gases, hydrogen and oxygen, can combine to form a liquid, water, in an exothermic reaction, as described by the following equation:

$$2\ H_2 + O_2 \rightarrow 2\ H_2O$$

Reaction stoichiometry describes the 2:1:2 ratio of hydrogen, oxygen, and water molecules in the above equation.

The molar ratio allows for conversion between moles of one substance and moles of another. For example, in the following reaction:

$$2\ CH_3OH + 3\ O_2 \rightarrow 2\ CO_2 + 4\ H_2O$$

the amount of water that will be produced by the combustion of 0.27 moles of CH3OH is obtained using the molar ratio between CH_3OH and H_2O of 2 to 4.

$$\left(\frac{0.27 \text{ mol } CH_3OH}{1}\right)\left(\frac{4 \text{ mol } H_2O}{2 \text{ mol } CH_3OH}\right) = 0.54 \text{ mol } H_2O$$

The term stoichiometry is also often used for the molar proportions of elements in stoichiometric compounds (composition stoichiometry). For example, the stoichiometry of hydrogen and oxygen in H_2O is 2:1. In stoichiometric compounds, the molar proportions are whole numbers.

Determining Amount of Product

Stoichiometry can also be used to find the quantity of a product yielded by a reaction. If a piece of solid copper (Cu) were added to an aqueous solution of silver nitrate ($AgNO_3$), the silver (Ag) would be replaced in a single displacement reaction forming aqueous copper(II) nitrate ($Cu(NO_3)_2$) and solid silver. How much silver is produced if 16.00 grams of Cu is added to the solution of excess silver nitrate?

The following steps would be used:

1. Write and balance the equation.
2. Mass to moles: Convert grams of Cu to moles of Cu.
3. Mole ratio: Convert moles of Cu to moles of Ag produced.
4. Mole to mass: Convert moles of Ag to grams of Ag produced.

The complete balanced equation would be:

$$Cu + 2\ AgNO_3 \rightarrow Cu(NO_3)2 + 2\ Ag$$

For the mass to mole step, the mass of copper (16.00 g) would be converted to moles of copper by dividing the mass of copper by its molecular mass: 63.55 g/mol.

$$\left(\frac{16.00 \text{ g Cu}}{1}\right)\left(\frac{1 \text{ mol Cu}}{63.55 \text{ g Cu}}\right) = 0.2518 \text{ mol Cu}$$

Now that the amount of Cu in moles (0.2518) is found, we can set up the mole ratio. This is found by looking at the coefficients in the balanced equation: Cu and Ag are in a 1:2 ratio.

$$\left(\frac{0.2518 \text{ mol Cu}}{1}\right)\left(\frac{2 \text{ mol Ag}}{1 \text{ mol Cu}}\right) = 0.5036 \text{ mol Ag}$$

Now that the moles of Ag produced is known to be 0.5036 mol, we convert this amount to grams of Ag produced to come to the final answer:

$$\left(\frac{0.5036 \text{ mol Ag}}{1}\right)\left(\frac{107.87 \text{ g Ag}}{1 \text{ mol Ag}}\right) = 54.32 \text{ g Ag}$$

This set of calculations can be further condensed into a single step:

$$\text{m}_{\text{Ag}} = \left(\frac{16.00 \text{ g Cu}}{1}\right)\left(\frac{1 \text{ mol Cu}}{63.55 \text{ g Cu}}\right)\left(\frac{2 \text{ mol Ag}}{1 \text{ mol Cu}}\right)\left(\frac{107.87 \text{ g Ag}}{1 \text{ mol Ag}}\right) = 54.32 \text{ g}$$

Further Examples

For propane (C_3H_8) reacting with oxygen gas (O_2), the balanced chemical equation is:

$$C_3H_8 + 5\ O_2 \rightarrow 3\ CO_2 + 4\ H_2O$$

The mass of water formed if 120 g of propane (C_3H_8) is burned in excess oxygen is then,

$$m_{H_2O} = \left(\frac{120.\text{ g } C_3H_8}{1}\right)\left(\frac{1 \text{ mol } C_3H_8}{44.09 \text{ g } C_3H_8}\right)\left(\frac{4 \text{ mol } H_2O}{1 \text{ mol } C_3H_8}\right)\left(\frac{18.02 \text{ g } H_2O}{1 \text{ mol } H_2O}\right) = 196 \text{ g}$$

Stoichiometric Ratio

Stoichiometry is also used to find the right amount of one reactant to "completely" react with the other reactant in a chemical reaction – that is, the stoichiometric amounts that would result in no leftover reactants when the reaction takes place. An example is shown below using the thermite reaction,

$$Fe_2O_3 + 2\ Al \rightarrow Al_2O_3 + 2\ Fe$$

This equation shows that 1 mole of iron(III) oxide and 2 moles of aluminum will produce 1 mole of aluminium oxide and 2 moles of iron. So, to completely react with 85.0 g of iron(III) oxide (0.532 mol), 28.7 g (1.06 mol) of aluminium are needed.

$$m_{Al} = \left(\frac{85.0 \text{ g } Fe_2O_3}{1}\right)\left(\frac{1 \text{ mol } Fe_2O_3}{159.7 \text{ g } Fe_2O_3}\right)\left(\frac{2 \text{ mol Al}}{1 \text{ mol } Fe_2O_3}\right)\left(\frac{26.98 \text{ g Al}}{1 \text{ mol Al}}\right) = 28.7 \text{ g}$$

Limiting Reagent and Percent Yield

The limiting reagent is the reagent that limits the amount of product that can be formed and is completely consumed when the reaction is complete. An excess reactant is a reactant that is left over once the reaction has stopped due to the limiting reactant being exhausted.

Consider the equation of roasting lead(II) sulfide (PbS) in oxygen (O_2) to produce lead(II) oxide (PbO) and sulfur dioxide (SO_2):

$$2\ PbS + 3\ O_2 \rightarrow 2\ PbO + 2\ SO_2$$

To determine the theoretical yield of lead(II) oxide if 200.0 g of lead(II) sulfide and 200.0 g of oxygen are heated in an open container:

$$m_{PbO} = \left(\frac{200.0\ g\ PbS}{1}\right)\left(\frac{1\ mol\ PbS}{239.27\ g\ PbS}\right)\left(\frac{2\ mol\ PbO}{2\ mol\ PbS}\right)\left(\frac{223.2\ g\ PbO}{1\ mol\ PbO}\right) = 186.6\ g$$

$$m_{PbO} = \left(\frac{200.0\ g\ O_2}{1}\right)\left(\frac{1\ mol\ O_2}{32.00\ g\ O_2}\right)\left(\frac{2\ mol\ PbO}{3\ mol\ O_2}\right)\left(\frac{223.2\ g\ PbO}{1\ mol\ PbO}\right) = 930.0\ g$$

Because a lesser amount of PbO is produced for the 200.0 g of PbS, it is clear that PbS is the limiting reagent.

In reality, the actual yield is not the same as the stoichiometrically-calculated theoretical yield. Percent yield, then, is expressed in the following equation:

$$\text{percent yield} = \frac{\text{actual yield}}{\text{theoretical yield}}$$

If 170.0 g of lead(II) oxide is obtained, then the percent yield would be calculated as follows:

$$\text{percent yield} = \frac{170.0\ g\ PbO}{186.6\ g\ PbO} = 91.12\%$$

Example

Consider the following reaction, in which iron(III) chloride reacts with hydrogen sulfide to produce iron(III) sulfide and hydrogen chloride:

$$2\ FeCl_3 + 3\ H_2S \rightarrow Fe_2S_3 + 6\ HCl$$

Suppose 90.0 g of $FeCl_3$ reacts with 52.0 g of H_2S. To find the limiting reagent and the mass of HCl produced by the reaction, we could set up the following equations:

$$m_{HCl} = \left(\frac{90.0\ g\ FeCl_3}{1}\right)\left(\frac{1\ mol\ FeCl_3}{162\ g\ FeCl_3}\right)\left(\frac{6\ mol\ HCl}{2\ mol\ FeCl_3}\right)\left(\frac{36.5\ g\ HCl}{1\ mol\ HCl}\right) = 60.8\ g$$

$$m_{HCl} = \left(\frac{52.0\ g\ H_2S}{1}\right)\left(\frac{1\ mol\ H_2S}{34.1\ g\ H_2S}\right)\left(\frac{6\ mol\ HCl}{3\ mol\ H_2S}\right)\left(\frac{36.5\ g\ HCl}{1\ mol\ HCl}\right) = 111\ g$$

Thus, the limiting reagent is $FeCl_3$ and the amount of HCl produced is 60.8 g.

To find what mass of excess reagent (H_2S) remains after the reaction, we would set up the calculation to find out how much H_2S reacts completely with the 90.0 g $FeCl_3$:

$$m_{H_2S} = \left(\frac{90.0 \text{ g FeCl}_3}{1}\right)\left(\frac{1 \text{ mol FeCl}_3}{162 \text{ g FeCl}_3}\right)\left(\frac{3 \text{ mol H}_2\text{S}}{2 \text{ mol FeCl}_3}\right)\left(\frac{34.1 \text{ g H}_2\text{S}}{1 \text{ mol H}_2\text{S}}\right) = 28.4 \text{ g reacted}$$

By subtracting this amount from the original amount of H_2S, we can come to the answer:

$$52.0 \text{ g H}_2\text{S} - 28.4 \text{ g H}_2\text{S} = 23.6 \text{ g H}_2\text{S excess}$$

Different Stoichiometries in Competing Reactions

Often, more than one reaction is possible given the same starting materials. The reactions may differ in their stoichiometry. For example, the methylation of benzene (C_6H_6), through a Friedel–Crafts reaction using $AlCl_3$ as a catalyst, may produce singly methylated ($C_6H_5CH_3$), doubly methylated ($C_6H_4(CH_3)_2$), or still more highly methylated ($C_6H_{6-n}(CH_3)_n$) products, as shown in the following example,

$$C_6H_6 + CH_3Cl \rightarrow C_6H_5CH_3 + HCl$$

$$C_6H_6 + 2\ CH_3Cl \rightarrow C_6H_4(CH_3)_2 + 2\ HCl$$

$$C_6H_6 + n\ CH_3Cl \rightarrow C_6H_{6-n}(CH_3)_n + n\ HCl$$

In this example, which reaction takes place is controlled in part by the relative concentrations of the reactants.

Stoichiometric Coefficient

In lay terms, the *stoichiometric coefficient* (or *stoichiometric number* in the IUPAC nomenclature) of any given component is the number of molecules that participate in the reaction as written.

For example, in the reaction $CH_4 + 2\ O_2 \rightarrow CO_2 + 2\ H_2O$, the stoichiometric coefficient of CH_4 is −1, the stoichiometric coefficient of O_2 is −2, for CO_2 it would be +1 and for H_2O it is +2.

In more technically precise terms, the stoichiometric coefficient in a chemical reaction-system of the *i*th component is defined as,

$$\nu_i = \frac{\Delta N_i}{\Delta \xi}$$

or

$$\Delta N_i = \nu_i \Delta \xi$$

where N_i is the number of molecules of *i*, and ξ is the progress variable or extent of reaction.

The extent of reaction ξ can be regarded as [the amount of] a real (or hypothetical) product, one molecule of which produced each time the reaction event occurs. It is the extensive quantity describing the progress of a chemical reaction equal to the number of chemical transformations, as indicated by the reaction equation on a molecular scale, divided by the Avogadro constant (in essence, it is the amount of chemical transformations). The change in the extent of reaction is given by $d\xi = dn_B/\nu_B$, where ν_B is the stoichiometric number of any reaction entity B (reactant or product) and n_B is the corresponding amount.

The stoichiometric coefficient ν_i represents the degree to which a chemical species participates in a reaction. The convention is to assign negative coefficients to *reactants* (which are consumed) and positive ones to *products*. However, any reaction may be viewed as going in the reverse direction, and all the coefficients then change sign (as does the free energy). Whether a reaction actually *will* go in the arbitrarily selected forward direction or not depends on the amounts of the substances present at any given time, which determines the kinetics and thermodynamics, i.e., whether equilibrium lies to the *right* or the *left*.

In reaction mechanisms, stoichiometric coefficients for each step are always integers, since elementary reactions always involve whole molecules. If one uses a composite representation of an overall reaction, some may be rationalfractions. There are often chemical species present that do not participate in a reaction; their stoichiometric coefficients are therefore zero. Any chemical species that is regenerated, such as a catalyst, also has a stoichiometric coefficient of zero.

The simplest possible case is an isomerization:

$$A \rightarrow B$$

in which $\nu_B = 1$ since one molecule of B is produced each time the reaction occurs, while $\nu_A = -1$ since one molecule of A is necessarily consumed. In any chemical reaction, not only is the total mass conserved but also the numbers of atoms of each kind are conserved, and this imposes corresponding constraints on possible values for the stoichiometric coefficients.

There are usually multiple reactions proceeding simultaneously in any natural reaction system, including those in biology. Since any chemical component can participate in several reactions simultaneously, the stoichiometric coefficient of the *i*th component in the *k*th reaction is defined as:

$$v_{ik} = \frac{\partial N_i}{\partial \xi_k}$$

so that the total (differential) change in the amount of the *i*th component is:

$$dN_i = \sum_k \nu_{ik} d\xi_k .$$

Extents of reaction provide the clearest and most explicit way of representing compositional change, although they are not yet widely used.

With complex reaction systems, it is often useful to consider both the representation of a reaction system in terms of the amounts of the chemicals present $\{ N_i \}$ (state variables), and the representation in terms of the actual compositional degrees of freedom, as expressed by the extents of reaction $\{ \xi_k \}$. The transformation from a vector expressing the extents to a vector expressing the amounts uses a rectangular matrix whose elements are the stoichiometric coefficients $[\nu_{ik}]$.

The maximum and minimum for any ξ_k occur whenever the first of the reactants is depleted for the forward reaction; or the first of the "products" is depleted if the reaction as viewed as being pushed in the reverse direction. This is a purely kinematic restriction on the reaction simplex, a hyperplane in composition space, or *N*space, whose dimensionality equals the number of *linearly-independent* chemical reactions. This is necessarily less than the number of chemical components, since each reaction manifests a relation between at least two chemicals. The accessible region of the hyperplane depends on the amounts of each chemical species actually present, a contingent fact. Different such amounts can even generate different hyperplanes, all sharing the same algebraic stoichiometry.

In accord with the principles of chemical kinetics and thermodynamic equilibrium, every chemical reaction is *reversible*, at least to some degree, so that each equilibrium point must be an interior point of the simplex. As a consequence, extrema for the ξs will not occur unless an experimental system is prepared with zero initial amounts of some products.

The number of *physically*-independent reactions can be even greater than the number of chemical components, and depends on the various reaction mechanisms. For example, there may be two (or more) reaction *paths* for the isomerism above. The reaction may occur by itself, but faster and with different intermediates, in the presence of a catalyst.

The (dimensionless) "units" may be taken to be molecules or moles. Moles are most commonly used, but it is more suggestive to picture incremental chemical reactions in terms of molecules. The *N*s and ξs are reduced to molar units by dividing by Avogadro's number. While dimensional mass units may be used, the comments about integers are then no longer applicable.

Stoichiometry Matrix

In complex reactions, stoichiometries are often represented in a more compact form called the stoichiometry matrix. The stoichiometry matrix is denoted by the symbol N.

If a reaction network has n reactions and m participating molecular species then the stoichiometry matrix will have correspondingly m rows and n columns.

For example, consider the system of reactions shown below:

$$S_1 \rightarrow S_2$$

$$5\ S_3 + S_2 \rightarrow 4\ S_3 + 2\ S_2$$

$$S_3 \rightarrow S_4$$

$$S_4 \rightarrow S_5$$

This systems comprises four reactions and five different molecular species. The stoichiometry matrix for this system can be written as:

$$\mathbf{N} = \begin{bmatrix} -1 & 0 & 0 & 0 \\ 1 & 1 & 0 & 0 \\ 0 & -1 & -1 & 0 \\ 0 & 0 & 1 & -1 \\ 0 & 0 & 0 & 1 \end{bmatrix}$$

where the rows correspond to S_1, S_2, S_3, S_4 and S_5, respectively. Note that the process of converting a reaction scheme into a stoichiometry matrix can be a lossy transformation, for example, the stoichiometries in the second reaction simplify when included in the matrix. This means that it is not always possible to recover the original reaction scheme from a stoichiometry matrix.

Often the stoichiometry matrix is combined with the rate vector, v, and the species vector, S to form a compact equation describing the rates of change of the molecular species:

$$\frac{d\mathbf{S}}{dt} = \mathbf{N} \cdot \mathbf{v}.$$

Gas Stoichiometry

Gas stoichiometry is the quantitative relationship (ratio) between reactants and products in a chemical reaction with reactions that produce gases. Gas stoichiometry applies when the gases produced are assumed to be ideal, and the temperature, pressure, and volume of the gases are all known. The ideal gas law is used for these calculations. Often, but not always, the standard temperature and pressure (STP) are taken as 0 °C and 1 bar and used as the conditions for gas stoichiometric calculations.

Gas stoichiometry calculations solve for the unknown volume or mass of a gaseous product or reactant. For example, if we wanted to calculate the volume of gaseous NO_2 produced from the combustion of 100 g of NH_3, by the reaction:

$$4\ NH_3(g) + 7\ O_2(g) \rightarrow 4\ NO_2(g) + 6\ H_2O(l)$$

we would carry out the following calculations:

$$100 g NH_3 \cdot \frac{1 mol NH_3}{17.034 g NH_3} = 5.871 mol NH_3$$

There is a 1:1 molar ratio of NH_3 to NO_2 in the above balanced combustion reaction, so 5.871 mol of NO_2 will be formed. We will employ the ideal gas law to solve for the volume at 0 °C (273.15 K) and 1 atmosphere using the gas law constant of $R = 0.08206$ $L \cdot atm \cdot K^{-1} \cdot mol^{-1}$:

$$PV = nRT$$

$$V = \frac{nRT}{P}$$

$$= \frac{5.871 \cdot 0.08206 \cdot 273.15}{1}$$

$$= 131.597 L NO_2$$

Gas stoichiometry often involves having to know the molar mass of a gas, given the density of that gas. The ideal gas law can be re-arranged to obtain a relation between the density and the molar mass of an ideal gas:

$$\rho = \frac{m}{V} \quad \text{and} \quad n = \frac{m}{M}$$

and thus:

$$\rho = \frac{MP}{RT}$$

where:

- P = absolute gas pressure
- V = gas volume
- n = amount (measured in moles)
- R = universal ideal gas law constant
- T = absolute gas temperature

- ρ = gas density at T and P
- m = mass of gas
- M = molar mass of gas

Stoichiometric air-to-fuel Ratios of Common Fuels

In the combustion reaction, oxygen reacts with the fuel, and the point where exactly all oxygen is consumed and all fuel burned is defined as the stoichiometric point. With more oxygen (overstoichiometric combustion), some of it stays unreacted. Likewise, if the combustion is incomplete due to lack of sufficient oxygen, fuel remains unreacted. (Unreacted fuel may also remain because of slow combustion or insufficient mixing of fuel and oxygen – this is not due to stoichiometry). Different hydrocarbon fuels have different contents of carbon, hydrogen and other elements, thus their stoichiometry varies.

Fuel	Ratio by mass	Ratio by volume	Percent fuel by mass
Gasoline	14.7 : 1	—	6.8%
Natural gas	17.2 : 1	9.7 : 1	5.8%
Propane (LP)	15.67 : 1	23.9 : 1	6.45%
Ethanol	9 : 1	—	11.1%
Methanol	6.47 : 1	—	15.6%
n-Butanol	11.2 : 1	—	8.2%
Hydrogen	34.3 : 1	2.39 : 1	2.9%
Diesel	14.5 : 1	—	6.8%
Methane	17.19 : 1	9.52 : 1	5.5%

Gasoline engines can run at stoichiometric air-to-fuel ratio, because gasoline is quite volatile and is mixed (sprayed or carburetted) with the air prior to ignition. Diesel engines, in contrast, run lean, with more air available than simple stoichiometry would require. Diesel fuel is less volatile and is effectively burned as it is injected, leaving less time for evaporation and mixing. Thus, it would form soot (black smoke) at stoichiometric ratio.

Gas Constant

Values of *R*	Units ($VPT^{-1}n^{-1}$)
8.314 4598(48)	kg m^2 s^{-2} K^{-1} mol^{-1}
8.3144598(48)	J K^{-1} mol^{-1}
	kJ K^{-1} $kmol^{-1}$
8.3144598(48)×10^7	erg K^{-1} mol^{-1}

$8.3144598(48)\times10^{-3}$	amu $(km/s)^2 K^{-1}$
8.3144598(48)	$m^3 Pa K^{-1} mol^{-1}$
$8.3144598(48)\times10^{+6}$	$cm^3 Pa K^{-1} mol^{-1}$
8.3144598(48)	L kPa $K^{-1} mol^{-1}$
$8.3144598(48)\times10^{3}$	cm^3 kPa $K^{-1} mol^{-1}$
$8.3144598(48)\times10^{-6}$	m^3 MPa $K^{-1} mol^{-1}$
8.3144598(48)	cm^3 MPa $K^{-1} mol^{-1}$
$8.3144598(48)\times10^{-5}$	m^3 bar $K^{-1} mol^{-1}$
$8.3144598(48)\times10^{-2}$	L bar $K^{-1} mol^{-1}$
83.144598(48)	cm^3 bar $K^{-1} mol^{-1}$
62.363577(36)	L Torr $K^{-1} mol^{-1}$
$1.9872036(11)\times10^{-3}$	kcal $K^{-1} mol^{-1}$
$8.2057338(47)\times10^{-5}$	m^3 atm $K^{-1} mol^{-1}$
$8.2057338(47)\times10^{-2}$	L atm $K^{-1} mol^{-1}$
82.057338(47)	cm^3 atm $K^{-1} mol^{-1}$

The gas constant (also known as the molar, universal, or ideal gas constant, denoted by the symbol *R* or *R*) is a physical constant which is featured in many fundamental equations in the physical sciences, such as the ideal gas law and the Nernst equation.

It is equivalent to the Boltzmann constant, but expressed in units of energy (i.e. the pressure-volume product) per temperature increment per *mole* (rather than energy per temperature increment per *particle*). The constant is also a combination of the constants from Boyle's law, Charles's law, Avogadro's law, and Gay-Lussac's law.

Physically, the gas constant is the constant of proportionality that happens to relate the energy scale in physics to the temperature scale, when a mole of particles at the stated temperature is being considered. Thus, the value of the gas constant ultimately derives from historical decisions and accidents in the setting of the energy and temperature scales, plus similar historical setting of the value of the molar scale used for the counting of particles. The last factor is not a consideration in the value of the Boltzmann constant, which does a similar job of equating linear energy and temperature scales.

The gas constant value is:

$$8.3144598(48)\ J\cdot mol^{-1}\cdot K^{-1}$$

The two digits in parentheses are the uncertainty (standard deviation) in the last two digits of the value. The relative uncertainty is 5.7×10^{-7}. Some have suggested that it might be appropriate to name the symbol *R* the Regnault constant in honour of the FrenchchemistHenri Victor Regnault, whose accurate experimental data were used to calculate the early value of the constant; however, the exact reason for the original representation of the constant by the letter *R* is elusive.

The gas constant occurs in the ideal gas law, as follows:

$$PV = nRT = mR_{specific}T$$

where *P* is the absolute pressure (SI unit pascals), *V* is the volume of gas (SI unit cubic metres), *n* is the amount of gas (SI unit moles), *m* is the mass (SI unit kilograms) contained in *V*, and *T* is the thermodynamic temperature (SI unit kelvin). The gas constant is expressed in the same physical units as molar entropy and molar heat capacity.

Dimensions of *R*

From the general equation $PV = nRT$ we get:

$$R = \frac{PV}{nT}$$

where *P* is pressure, *V* is volume, *n* is number of moles of a given substance, and *T* is temperature.

As pressure is defined as force per unit area, the gas equation can also be written as:

$$R = \frac{\frac{\text{force}}{\text{area}} \times \text{volume}}{\text{amount} \times \text{temperature}}$$

Area and volume are $(\text{length})^2$ and $(\text{length})^3$ respectively. Therefore:

$$R = \frac{\frac{\text{force}}{(\text{length})^2} \times (\text{length})^3}{\text{amount} \times \text{temperature}} = \frac{\text{force} \times \text{length}}{\text{amount} \times \text{temperature}}$$

Since force × length = work:

$$R = \frac{\text{work}}{\text{amount} \times \text{temperature}}$$

The physical significance of *R* is work per degree per mole. It may be expressed in any set of units representing work or energy (such as joules), other units representing degrees of temperature (such as degrees Celsius or Fahrenheit), and any system of units designating a mole or a similar pure number that allows an equation of macroscopic mass and fundamental particle numbers in a system, such as an ideal gas.

Instead of a mole the constant can be expressed by considering the normal cubic meter.

Otherwise, we can also say that:

$$\text{force} = \frac{\text{mass} \times \text{length}}{(\text{time})^2}$$

Therefore, we can write "R" as:

$$R = \frac{\text{mass} \times \text{length}^2}{\text{amount} \times \text{temperature} \times (\text{time})^2}$$

And so, in SI base units:

$$R = 8.3144598(48)\ \text{kg m}^2\ \text{mol}^{-1}\ \text{K}^{-1}\ \text{s}^{-2}.$$

Relationship with the Boltzmann Constant

The Boltzmann constantk_B (often abbreviated k) may be used in place of the gas constant by working in pure particle count, N, rather than amount of substance, n, since:

$$R = N_A k_B,$$

where N_A is the Avogadro constant. For example, the ideal gas law in terms of Boltzmann's constant is,

$$PV = k_B NT.$$

where N is the number of particles (molecules in this case), or to generalize to an inhomogeneous system the local form holds:

$$P = k_B nT.$$

where n is the number density.

Measurement

As of 2006, the most precise measurement of R is obtained by measuring the speed of sound $c_a(p, T)$ in argon at the temperature T of the triple point of water (used to define the kelvin) at different pressures p, and extrapolating to the zero-pressure limit $c_a(0, T)$. The value of R is then obtained from the relation:

$$c_a(0,T) = \sqrt{\frac{\gamma_0 RT}{A_r(\text{Ar})M_u}},$$

where:

- γ_0 is the heat capacity ratio ($5/3$ for monatomic gases such as argon);
- T is the temperature, $T_{TPW} = 273.16$ K by definition of the kelvin;
- $A_r(\text{Ar})$ is the relative atomic mass of argon and $M_u = 10^{-3}$ kg mol^{-1}.

Specific Gas Constant

$R_{specific}$ for dry air	Units
287.058	J kg^{-1} K^{-1}
53.3533	ft lbf lb^{-1} °R^{-1}
1,716.49	ft lbf slug^{-1} °R^{-1}
Based on a mean molar mass for dry air of 28.9645 g/mol.	

The specific gas constant of a gas or a mixture of gases ($R_{specific}$) is given by the molar gas constant divided by the molar mass (M) of the gas or mixture.

$$R_{specific} = \frac{R}{M}$$

Just as the ideal gas constant can be related to the Boltzmann constant, so can the specific gas constant by dividing the Boltzmann constant by the molecular mass of the gas.

$$R_{specific} = \frac{k_B}{m}$$

Another important relationship comes from thermodynamics. Mayer's relation relates the specific gas constant to the specific heats for a calorically perfect gas and a thermally perfect gas.

$$R_{specific} = c_p - c_v$$

where c_p is the specific heat for a constant pressure and c_v is the specific heat for a constant volume.

It is common, especially in engineering applications, to represent the specific gas constant by the symbol R. In such cases, the universal gas constant is usually given a different symbol such as $\bar{R}$ to distinguish it. In any case, the context and/or units of the gas constant should make it clear as to whether the universal or specific gas constant is being referred to.

U.S. Standard Atmosphere

The U.S. Standard Atmosphere, 1976 (USSA1976) defines the gas constant R^* as:

$$R^* = 8.31432 \times 10^3 \text{ N m kmol}^{-1} \text{ K}^{-1}.$$

Note the use of kilomole units resulting in the factor of 1,000 in the constant. The USSA1976 acknowledges that this value is not consistent with the cited values for the Avogadro constant and the Boltzmann constant. This disparity is not a significant departure from accuracy, and USSA1976 uses this value of R^* for all the calculations of the standard atmosphere. When using the ISO value of R, the calculated pressure increases by only 0.62 pascal at 11 kilometers (the equivalent of a difference of only 17.4 centimeters or 6.8 inches) and an increase of 0.292 Pa at 20 km (the equivalent of a difference of only 0.338 m or 13.2 in).

Excess and Limiting Reactants

The exact amount of the reactants can be determined by the stoichiometry of the reaction. In many cases one or more reactants is supplied in excess to complete the reaction. Reactant present in the smallest stoichiometric amount is called the limiting reactant. Reactants supplied in excess of the limiting reactant are called excess reactants.

Oxidation- Reduction Reactions

Oxidation: Loss of electrons from an atom.

Reduction: Gain of electrons by an atom.

Oxidation and reduction always occurs simultaneously so that electrons are conserved.

Cementation reaction.

Thermochemistry

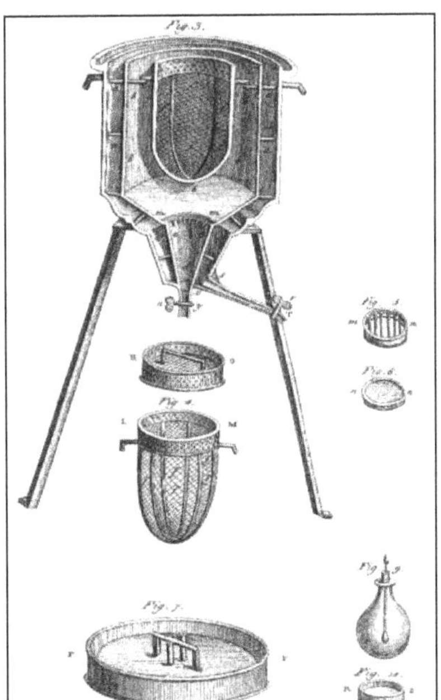

The world's first ice-calorimeter, used in the winter of 1782-83, by Antoine Lavoisier and Pierre-Simon Laplace, to determine the heat evolved in various chemical changes; calculations which were based on Joseph Black's prior discovery of latent heat. These experiments mark the foundation of thermochemistry.

Thermochemistry is the study of the energy and heat associated with chemical reactions and/or physical transformations. A reaction may release or absorb energy, and a phase change may do the same, such as in melting and boiling. Thermochemistry focuses on these energy changes, particularly on the system's energy exchange with its surroundings. Thermochemistry is useful in predicting reactant and product quantities throughout the course of a given reaction. In combination with entropy determinations, it is also used to predict whether a reaction is spontaneous or non-spontaneous, favorable or unfavorable.

Endothermic reactions absorb heat, while exothermic reactions release heat. Thermochemistry coalesces the concepts of thermodynamics with the concept of energy in the form of chemical bonds. The subject commonly includes calculations of such quantities as heat capacity, heat of combustion, heat of formation, enthalpy, entropy, free energy, and calories.

History

Thermochemistry rests on two generalizations. Stated in modern terms, they are as follows:

1. Lavoisier and Laplace's law (1780): The energy change accompanying any transformation is equal and opposite to energy change accompanying the reverse process.

2. Hess' law (1840): The energy change accompanying any transformation is the same whether the process occurs in one step or many.

These statements preceded the first law of thermodynamics (1845) and helped in its formulation.

Lavoisier, Laplace and Hess also investigated specific heat and latent heat, although it was Joseph Black who made the most important contributions to the development of latent energy changes.

Gustav Kirchhoff showed in 1858 that the variation of the heat of reaction is given by the difference in heat capacity between products and reactants: $d\Delta H / dT = \Delta C_p$. Integration of this equation permits the evaluation of the heat of reaction at one temperature from measurements at another temperature.

Calorimetry

The measurement of heat changes is performed using calorimetry, usually an enclosed chamber within which the change to be examined occurs. The temperature of the chamber is monitored either using a thermometer or thermocouple, and the temperature plotted against time to give a graph from which fundamental quantities can be calculated.

Modern calorimeters are frequently supplied with automatic devices to provide a quick read-out of information, one example being the differential scanning calorimeter (DSC).

Systems

Several thermodynamic definitions are very useful in thermochemistry. A system is the specific portion of the universe that is being studied. Everything outside the system is considered the surroundings or environment. A system may be:

- a (completely) isolated system which can exchange neither energy nor matter with the surroundings, such as an insulated bomb calorimeter.
- a thermally isolated system which can exchange mechanical work but not heat or matter, such as an insulated closed piston or balloon.
- a mechanically isolated system which can exchange heat but not mechanical work or matter, such as an uninsulated bomb calorimeter.
- a closed system which can exchange energy but not matter, such as an uninsulated closed piston or balloon.
- an open system which it can exchange both matter and energy with the surroundings, such as a pot of boiling water.

Processes

A system undergoes a process when one or more of its properties changes. A process relates to the change of state. An isothermal (same-temperature) process occurs when temperature of the system remains constant. An isobaric (same-pressure) process occurs when the pressure of the system remains constant. A process is adiabatic when no heat exchange occurs.

First Law of Thermodynamics

The first law of thermodynamics is a version of the law of conservation of energy, adapted for thermodynamic systems. The law of conservation of energy states that the total energy of an isolated system is constant; energy can be transformed from one form to another, but can be neither created nor destroyed. The first law is often formulated by stating that the change in the internal energy of a closed system is equal to the amount of heat supplied to the system, minus the amount of work done by the system on its surroundings. Equivalently, perpetual motion machines of the first kind are impossible.

History

Investigations into the nature of heat and work and their relationship began with the invention of the first engines used to extract water from mines. Improvements to such engines so as to increase their efficiency and power output came first from mechanics that worked with such machines but only slowly advanced the art. Deeper investigations that placed those on a mathematical and physics basis came later.

The first law of thermodynamics was developed empirically over about half a century. The first full statements of the law came in 1850 from Rudolf Clausius and from William Rankine; Rankine's statement is less distinct relative to Clausius'. A main aspect of the struggle was to deal with the previously proposed caloric theory of heat.

In 1840, Germain Hess stated a conservation law for the so-called 'heat of reaction' for chemical reactions. His law was later recognized as a consequence of the first law of thermodynamics, but Hess's statement was not explicitly concerned with the relation between energy exchanges by heat and work.

According to Truesdell (1980), Julius Robert von Mayer in 1841 made a statement that meant that "in a process at constant pressure, the heat used to produce expansion is universally interconvertible with work", but this is not a general statement of the first law.

Original Statements: the Thermodynamic Approach

The original nineteenth century statements of the first law of thermodynamics appeared in a conceptual framework in which transfer of energy as heat was taken as a primitive notion, not defined or constructed by the theoretical development of the framework, but rather presupposed as prior to it and already accepted. The primitive notion of heat was taken as empirically established, especially through calorimetry regarded as a subject in its own right, prior to thermodynamics. Jointly primitive with this notion of heat were the notions of empirical temperature and thermal equilibrium. This framework also took as primitive the notion of transfer of energy as work. This framework did not presume a concept of energy in general, but regarded it as derived or synthesized from the prior notions of heat and work. By one author, this framework has been called the "thermodynamic" approach.

The first explicit statement of the first law of thermodynamics, by Rudolf Clausius in 1850, referred to cyclic thermodynamic processes.

> *In all cases in which work is produced by the agency of heat, a quantity of heat is consumed which is proportional to the work done; and conversely, by the expenditure of an equal quantity of work an equal quantity of heat is produced.*

Clausius also stated the law in another form, referring to the existence of a function of state of the system, the internal energy, and expressed it in terms of a differential equation for the increments of a thermodynamic process. This equation may be described as follows:

In a thermodynamic process involving a closed system, the increment in the internal energy is equal to the difference between the heat accumulated by the system and the work done by it.

Because of its definition in terms of increments, the value of the internal energy of a system is not uniquely defined. It is defined only up to an arbitrary additive constant of integration, which can be adjusted to give arbitrary reference zero levels. This non-uniqueness is in keeping with the abstract mathematical nature of the internal energy. The internal energy is customarily stated relative to a conventionally chosen standard reference state of the system.

The concept of internal energy is considered by Bailyn to be of "enormous interest". Its quantity cannot be immediately measured, but can only be inferred, by differencing actual immediate measurements. Bailyn likens it to the energy states of an atom, that were revealed by Bohr's energy relation $h\nu = E_{n''} - E_{n'}$. In each case, an unmeasurable quantity (the internal energy, the atomic energy level) is revealed by considering the difference of measured quantities (increments of internal energy, quantities of emitted or absorbed radiative energy).

Conceptual Revision: the Mechanical Approach

In 1907, George H. Bryan wrote about systems between which there is no transfer of matter (closed systems): "Definition. When energy flows from one system or part of a system to another otherwise than by the performance of mechanical work, the energy so transferred is called *heat*." This definition may be regarded as expressing a conceptual revision, as follows. This was systematically expounded in 1909 by Constantin Carathéodory, whose attention had been drawn to it by Max Born. Largely through Born's influence, this revised conceptual approach to the definition of heat came to be preferred by many twentieth-century writers. It might be called the "mechanical approach".

Energy can also be transferred from one thermodynamic system to another in association with transfer of matter. Born points out that in general such energy transfer is not resolvable uniquely into work and heat moieties. In general, when there is transfer of energy associated with matter transfer, work and heat transfers can be distinguished only when they pass through walls physically separate from those for matter transfer.

The "mechanical" approach postulates the law of conservation of energy. It also postulates that energy can be transferred from one thermodynamic system to another adiabatically as work, and that energy can be held as the internal energy of a thermodynamic system. It also postulates that energy can be transferred from one thermodynamic system to another by a path that is non-adiabatic, and is unaccompanied by matter transfer. Initially, it "cleverly" (according to Bailyn) refrains from labelling as 'heat' such non-adiabatic, unaccompanied transfer of energy. It rests on the primitive notion of *walls*, especially adiabatic walls and non-adiabatic walls, defined as follows. Temporarily, only for purpose of this definition, one can prohibit transfer of energy as work across a wall of interest. Then walls of interest fall into two classes, (a) those such that arbitrary systems separated by them remain independently in their own previously established respective states of internal thermodynamic equilibrium; they are defined as adiabatic; and (b) those without such independence; they are defined as non-adiabatic.

This approach derives the notions of transfer of energy as heat, and of temperature, as theoretical developments, not taking them as primitives. It regards calorimetry as a derived theory. It has an early origin in the nineteenth century, for example in the work of Helmholtz, but also in the work of many others.

Conceptually Revised Statement: According to the Mechanical Approach

The revised statement of the first law postulates that a change in the internal energy of a system due to any arbitrary process, that takes the system from a given initial thermodynamic state to a given final equilibrium thermodynamic state, can be determined through the physical existence, for those given states, of a reference process that occurs purely through stages of adiabatic work.

The revised statement is then:

> *For a closed system, in any arbitrary process of interest that takes it from an initial to a final state of internal thermodynamic equilibrium, the change of internal energy is the same as that for a reference adiabatic work process that links those two states. This is so regardless of the path of the process of interest, and regardless of whether it is an adiabatic or a non-adiabatic process. The reference adiabatic work process may be chosen arbitrarily from amongst the class of all such processes.*

This statement is much less close to the empirical basis than are the original statements, but is often regarded as conceptually parsimonious in that it rests only on the concepts of adiabatic work and of non-adiabatic processes, not on the concepts of transfer of energy as heat and of empirical temperature that are presupposed by the original statements. Largely through the influence of Max Born, it is often regarded as theoretically preferable because of this conceptual parsimony. Born particularly observes that the revised approach avoids thinking in terms of what he calls the "imported engineering" concept of heat engines.

Basing his thinking on the mechanical approach, Born in 1921, and again in 1949, proposed to revise the definition of heat. In particular, he referred to the work of Constantin Carathéodory, who had in 1909 stated the first law without defining quantity of heat. Born's definition was specifically for transfers of energy without transfer of matter, and it has been widely followed in textbooks (examples:). Born observes that a transfer of matter between two systems is accompanied by a transfer of internal energy that cannot be resolved into heat and work components. There can be pathways to other systems, spatially separate from that of the matter transfer, that allow heat and work transfer independent of and simultaneous with the matter transfer. Energy is conserved in such transfers.

Description

The first law of thermodynamics for a closed system was expressed in two ways by Clausius. One way referred to cyclic processes and the inputs and outputs of the system, but did not refer to increments in the internal state of the system. The other way referred to an incremental change in the internal state of the system, and did not expect the process to be cyclic.

A cyclic process is one that can be repeated indefinitely often, returning the system to its initial state. Of particular interest for single cycle of a cyclic process are the net work done, and the net heat taken in (or 'consumed', in Clausius' statement), by the system.

In a cyclic process in which the system does net work on its surroundings, it is observed to be physically necessary not only that heat be taken into the system, but also, importantly, that some heat leave the system. The difference is the heat converted by the cycle into work. In each repetition of a cyclic process, the net work done by the system, measured in mechanical units, is proportional to the heat consumed, measured in calorimetric units.

The constant of proportionality is universal and independent of the system and in 1845 and 1847 was measured by James Joule, who described it as the *mechanical equivalent of heat*.

In a non-cyclic process, the change in the internal energy of a system is equal to net energy added as heat to the system minus the net work done by the system, both being measured in mechanical units. Taking ΔU as a change in internal energy, one writes:

$$\Delta U = Q - W$$

where Q denotes the net quantity of heat supplied to the system by its surroundings and W denotes the net work done by the system. This sign convention is implicit in Clausius' statement of the law given above. It originated with the study of heat engines that produce useful work by consumption of heat.

Often nowadays, however, writers use the IUPAC convention by which the first law is formulated with work done on the system by its surroundings having a positive sign. With this now often used sign convention for work, the first law for a closed system may be written:

$$\Delta U = Q + W \text{ (sign convention of IUPAC).}$$

This convention follows physicists such as Max Planck, and considers all net energy transfers to the system as positive and all net energy transfers from the system as negative, irrespective of any use for the system as an engine or other device.

When a system expands in a fictive quasistatic process, the work done by the system on the environment is the product, $P\,dV$, of pressure, P, and volume change, dV, whereas the work done *on* the system is $-P\,dV$. Using either sign convention for work, the change in internal energy of the system is:

$$dU = \delta Q - PdV \text{ (quasi-static process)},$$

where δQ denotes the infinitesimal increment of heat supplied to the system from its surroundings.

Work and heat are expressions of actual physical processes of supply or removal of energy, while the internal energy U is a mathematical abstraction that keeps account of the exchanges of energy that befall the system. Thus the term heat for Q means "that amount of energy added or removed by conduction of heat or by thermal radiation", rather than referring to a form of energy within the system. Likewise, the term work energy for W means "that amount of energy gained or lost as the result of work". Internal energy is a property of the system whereas work done and heat supplied are not. A significant result of this distinction is that a given internal energy change ΔU can be achieved by, in principle, many combinations of heat and work.

Various Statements of the Law for Closed Systems

The law is of great importance and generality and is consequently thought of from several points of view. Most careful textbook statements of the law express it for closed systems. It is stated in several ways, sometimes even by the same author.

For the thermodynamics of closed systems, the distinction between transfers of energy as work and as heat is central. For the thermodynamics of open systems, such a distinction is beyond the scope, but some limited comments are made on it in the section below headed 'First law of thermodynamics for open systems'.

There are two main ways of stating a law of thermodynamics, physically or mathematically. They should be logically coherent and consistent with one another.

An example of a physical statement is that of Planck:

> It is in no way possible, either by mechanical, thermal, chemical, or other devices, to obtain perpetual motion, i.e. it is impossible to construct an engine which will work in a cycle and produce continuous work, or kinetic energy, from nothing.

This physical statement is restricted neither to closed systems nor to systems with states that are strictly defined only for thermodynamic equilibrium; it has meaning also for open systems and for systems with states that are not in thermodynamic equilibrium.

An example of a mathematical statement is that of Crawford (1963):

> For a given system we let ΔE^{kin} = large-scale mechanical energy, ΔE^{pot} = large-scale potential energy, and ΔE^{tot} = total energy. The first two quantities are specifiable in terms of appropriate mechanical variables, and by definition,
>
> $$E^{\text{tot}} = E^{\text{kin}} + E^{\text{pot}} + U.$$
>
> For any finite process, whether reversible or irreversible,
>
> $$\Delta E^{\text{tot}} = \Delta E^{\text{kin}} + \Delta E^{\text{pot}} + \Delta U.$$
>
> The first law in a form that involves the principle of conservation of energy more generally is,
>
> $$\Delta E^{\text{tot}} = Q + W.$$
>
> Here Q and W are heat and work added, with no restrictions as to whether the process is reversible, quasistatic, or irreversible.

This statement by Crawford, for W, uses the sign convention of IUPAC, not that of Clausius. Though it does not explicitly say so, this statement refers to closed systems, and to internal energy U defined for bodies in states of thermodynamic equilibrium, which possess well-defined temperatures.

The history of statements of the law for closed systems has two main periods, before and after the work of Bryan (1907), of Carathéodory (1909), and the approval of Carathéodory's work given by Born (1921). The earlier traditional versions of the law for closed systems are nowadays often considered to be out of date.

Carathéodory's celebrated presentation of equilibrium thermodynamics refers to closed systems, which are allowed to contain several phases connected by internal walls of various kinds of impermeability and permeability (explicitly including walls that are permeable only to heat). Carathéodory's 1909 version of the first law of thermodynamics was stated in an axiom which refrained from defining or mentioning temperature or quantity of heat transferred. That axiom stated that the internal energy of a phase in equilibrium is a function of state, that the sum of the internal energies of the phases is the total internal energy of the system, and that the value of the total internal energy of the system is changed by the amount of work done adiabatically on it, considering work as a form of energy. The article considered this statement to be an expression of the law of conservation of energy for such systems. This version is nowadays widely accepted as authoritative, but is stated in slightly varied ways by different authors.

Such statements of the first law for closed systems assert the existence of internal energy as a function of state defined in terms of adiabatic work. Thus heat is not defined calorimetrically or as due to temperature difference. It is defined as a residual difference between change of internal energy and work done on the system, when that work does not account for the whole of the change of internal energy and the system is not adiabatically isolated.

The 1909 Carathéodory statement of the law in axiomatic form does not mention heat or temperature, but the equilibrium states to which it refers are explicitly defined by variable sets that necessarily include "non-deformation variables", such as pressures, which, within reasonable restrictions, can be rightly interpreted as empirical temperatures, and the walls connecting the phases of the system are explicitly defined as possibly impermeable to heat or permeable only to heat.

According to Münster (1970), "A somewhat unsatisfactory aspect of Carathéodory's theory is that a consequence of the Second Law must be considered at this point [in the statement of the first law], i.e. that it is not always possible to reach any state 2 from any other state 1 by means of an adiabatic process." Münster instances that no adiabatic process can reduce the internal energy of a system at constant volume. Carathéodory's paper asserts that its statement of the first law corresponds exactly to Joule's experimental arrangement, regarded as an instance of adiabatic work. It does not point out that Joule's experimental arrangement performed essentially irreversible work, through friction of paddles in a liquid, or passage of electric current through a resistance inside the system, driven by motion of a coil and inductive heating, or by an external current source, which can access the system only by the passage of electrons, and so is not strictly adiabatic, because electrons are a form of matter, which cannot penetrate adiabatic walls. The paper goes on to base its main argument on the possibility of quasi-static adiabatic work, which is essentially reversible. The paper asserts that it will avoid reference to Carnot cycles, and then proceeds to base its argument on cycles of forward and backward quasi-static adiabatic stages, with isothermal stages of zero magnitude.

Sometimes the concept of internal energy is not made explicit in the statement.

Sometimes the existence of the internal energy is made explicit but work is not explicitly mentioned in the statement of the first postulate of thermodynamics. Heat supplied is then defined as the residual change in internal energy after work has been taken into account, in a non-adiabatic process.

A respected modern author states the first law of thermodynamics as "Heat is a form of energy", which explicitly mentions neither internal energy nor adiabatic work. Heat is defined as energy transferred by thermal contact with a reservoir, which has a temperature, and is generally so large that addition and removal of heat do not alter its temperature. A current student text on chemistry defines heat thus: "*heat* is the exchange of thermal energy between a system and its surroundings caused by a temperature difference." The author then explains how heat is defined or measured by calorimetry, in terms of heat capacity, specific heat capacity, molar heat capacity, and temperature.

A respected text disregards the Carathéodory's exclusion of mention of heat from the statement of the first law for closed systems, and admits heat calorimetrically defined along with work and internal energy. Another respected text defines heat exchange as determined by temperature difference, but also mentions that the Born (1921) version is "completely rigorous". These versions follow the traditional approach that is now considered out of date, exemplified by that of Planck.

Evidence for the First Law of Thermodynamics for Closed Systems

The first law of thermodynamics for closed systems was originally induced from empirically observed evidence, including calorimetric evidence. It is nowadays, however, taken to provide the definition of heat via the law of conservation of energy and the definition of work in terms of changes in the external parameters of a system. The original discovery of the law was gradual over a period of perhaps half a century or more, and some early studies were in terms of cyclic processes.

The following is an account in terms of changes of state of a closed system through compound processes that are not necessarily cyclic. This account first considers processes for which the first law is easily verified because of their simplicity, namely adiabatic processes(in which there is no transfer as heat) and adynamic processes (in which there is no transfer as work).

Adiabatic Processes

In an adiabatic process, there is transfer of energy as work but not as heat. For all adiabatic process that takes a system from a given initial state to a given final state, irrespective of how the work is done, the respective eventual total quantities of energy transferred as work are one and the same, determined just by the given initial and final states. The work done on the system is defined and measured by changes in mechanical or quasi-mechanical variables external to the system. Physically, adiabatic transfer of energy as work requires the existence of adiabatic enclosures.

For instance, in Joule's experiment, the initial system is a tank of water with a paddle wheel inside. If we isolate the tank thermally, and move the paddle wheel with a pulley and a weight, we can relate the increase in temperature with the distance descended by the mass. Next, the system is returned to its initial state, isolated again, and the same amount of work is done on the tank using different devices (an electric motor, a chemical battery, a spring,...). In every case, the amount of work can be measured independently. The return to the initial state is not conducted by doing adiabatic work on the system. The evidence shows that the final state of the water (in particular, its temperature and volume) is the same in every case. It is irrelevant if the work is electrical, mechanical, chemical,... or if done suddenly or slowly, as long as it is performed in an adiabatic way, that is to say, without heat transfer into or out of the system.

Evidence of this kind shows that to increase the temperature of the water in the tank, the qualitative kind of adiabatically performed work does not matter. No qualitative kind of adiabatic work has ever been observed to decrease the temperature of the water in the tank.

A change from one state to another, for example an increase of both temperature and volume, may be conducted in several stages, for example by externally supplied electrical work on a resistor in the body, and adiabatic expansion allowing the body to do work on the surroundings. It needs to be shown that the time order of the stages, and their relative magnitudes, does not affect the amount of adiabatic work that needs to be done for the change of state. According to one respected scholar: "Unfortunately, it does not seem that experiments of this kind have ever been carried out carefully. ... We must therefore admit that the statement which we have enunciated here, and which is equivalent to the first law of thermodynamics, is not well founded on direct experimental evidence." Another expression of this view is "... no systematic precise experiments to verify this generalization directly have ever been attempted."

This kind of evidence, of independence of sequence of stages, combined with the above-mentioned evidence, of independence of qualitative kind of work, would show the existence of an important state variable that corresponds with adiabatic work, but not that such a state variable represented a conserved quantity. For the latter, another step of evidence is needed, which may be related to the concept of reversibility, as mentioned below.

That important state variable was first recognized and denoted U by Clausius in 1850, but he did not then name it, and he defined it in terms not only of work but also of heat transfer in the same process. It was also independently recognized in 1850 by Rankine, who also denoted it U ; and in 1851 by Kelvin who then called it "mechanical energy", and later "intrinsic energy". In 1865, after some hestitation, Clausius began calling his state function U "energy". In 1882 it was named as the *internal energy* by Helmholtz. If only adiabatic processes were of interest, and heat could be ignored, the concept of internal energy would hardly arise or be needed. The relevant physics would be largely covered by the concept of potential energy, as was intended in the 1847 paper of Helmholtz on the principle of conservation of energy, though that did not deal with forces that cannot be described by a potential, and thus did not fully justify the principle. Moreover, that paper was critical of the early work of Joule that had by then been performed. A great merit of the internal energy concept is that it frees thermodynamics from a restriction to cyclic processes, and allows a treatment in terms of thermodynamic states.

In an adiabatic process, adiabatic work takes the system either from a reference state O with internal energy U(O) to an arbitrary one A with internal energy U(A), or from the state A to the state O:

$$U(A) = U(O) - W_{O \to A}^{\text{adiabatic}} \text{ or } U(O) = U(A) - W_{A \to O}^{\text{adiabatic}}$$

Except under the special, and strictly speaking, fictional, condition of reversibility, only one of the processes adiabatic, $O \to A$ or adiabatic, $A \to O$ is empirically feasible by a simple application of externally supplied work. The reason for this is given as the second law of thermodynamics and is not considered in this article.

The fact of such irreversibility may be dealt with in two main ways, according to different points of view:

- Since the work of Bryan (1907), the most accepted way to deal with it nowadays, followed by Carathéodory, is to rely on the previously established concept of quasi-static processes, as follows. Actual physical processes of transfer of energy as work are always at least to some degree irreversible. The irreversibility is often due to mechanisms known as dissipative, that transform bulk kinetic energy into internal energy. Examples are friction and viscosity. If the process is performed more slowly, the frictional or viscous dissipation is less. In the limit of infinitely slow performance, the dissipation tends to zero and then the limiting process, though fictional rather than actual, is notionally reversible, and is called quasi-static. Throughout the course of the fictional limiting quasi-static process, the internal intensive variables of the system are equal to the external intensive variables, those that describe the reactive forces exerted by the surroundings. This can be taken to justify the formula,

$$W_{A \to O}^{\text{adiabatic, quasi-static}} = -W_{O \to A}^{\text{adiabatic, quasi-static}}$$

- Another way to deal with it is to allow that experiments with processes of heat transfer to or from the system may be used to justify the formula above. Moreover, it deals to some extent with the problem of lack of direct experimental evidence that the time order of stages of a process does not matter in the determination of internal energy. This way does not provide theoretical purity in terms of adiabatic work processes, but is empirically feasible, and is in accord with experiments actually done, such as the Joule experiments mentioned just above, and with older traditions.

The formula above allows that to go by processes of quasi-static adiabatic work from the state A to the state B we can take a path that goes through the reference state O, since the quasi-static adiabatic work is independent of the path,

$$-W_{A \to B}^{\text{adiabatic,quasi-static}} = -W_{A \to O}^{\text{adiabatic,quasi-static}} - W_{O \to B}^{\text{adiabatic,quasi-static}}$$

$$= W_{O\to A}^{\text{adiabatic, quasi-static}} - W_{O\to B}^{\text{adiabatic, quasi-static}} = -\mathrm{U(A)} + \mathrm{U(B)} = \Delta \mathrm{U}$$

This kind of empirical evidence, coupled with theory of this kind, largely justifies the following statement:

> *For all adiabatic processes between two specified states of a closed system of any nature, the net work done is the same regardless the details of the process, and determines a state function called internal energy,* U ."

Adynamic Processes

A complementary observable aspect of the first law is about heat transfer. Adynamic transfer of energy as heat can be measured empirically by changes in the surroundings of the system of interest by calorimetry. This again requires the existence of adiabatic enclosure of the entire process, system and surroundings, though the separating wall between the surroundings and the system is thermally conductive or radiatively permeable, not adiabatic. A calorimeter can rely on measurement of sensible heat, which requires the existence of thermometers and measurement of temperature change in bodies of known sensible heat capacity under specified conditions; or it can rely on the measurement of latent heat, through measurement of masses of material that change phase, at temperatures fixed by the occurrence of phase changes under specified conditions in bodies of known latent heat of phase change. The calorimeter can be calibrated by adiabatically doing externally determined work on it. The most accurate method is by passing an electric current from outside through a resistance inside the calorimeter. The calibration allows comparison of calorimetric measurement of quantity of heat transferred with quantity of energy transferred as work. According to one textbook, "The most common device for measuring ΔU is an adiabatic bomb calorimeter." According to another textbook, "Calorimetry is widely used in present day laboratories." According to one opinion, "Most thermodynamic data come from calorimetry..." According to another opinion, "The most common method of measuring "heat" is with a calorimeter."

When the system evolves with transfer of energy as heat, without energy being transferred as work, in an adynamic process, the heat transferred to the system is equal to the increase in its internal energy:

$$\mathrm{Q}_{\mathrm{A}\to\mathrm{B}}^{\text{adynamic}} = \Delta \mathrm{U}$$

General Case for Reversible Processes

Heat transfer is practically reversible when it is driven by practically negligibly small temperature gradients. Work transfer is practically reversible when it occurs so slowly that there are no frictional effects within the system; frictional effects outside the system should also be zero if the process is to be globally reversible. For a particular reversible process in general, the work done reversibly on the system, $W_{A\to B}^{\text{path } P_0,\text{reversible}}$, and the heat transferred reversibly to the system, $Q_{A\to B}^{\text{path } P_0,\text{reversible}}$ are not required to occur respectively adiabatically or adynamically, but they must belong to the same particular process defined by its particular reversible path, P_0, through the space of thermodynamic states. Then the work and heat transfers can occur and be calculated simultaneously.

Putting the two complementary aspects together, the first law for a particular reversible process can be written,

$$-W_{A\to B}^{\text{path } P_0,\text{reversible}} + Q_{A\to B}^{\text{path } P_0,\text{reversible}} = \Delta U$$

This combined statement is the expression the first law of thermodynamics for reversible processes for closed systems.

In particular, if no work is done on a thermally isolated closed system we have:

$$\Delta U = 0$$

This is one aspect of the law of conservation of energy and can be stated:

The internal energy of an isolated system remains constant.

General Case for Irreversible Processes

If, in a process of change of state of a closed system, the energy transfer is not under a practically zero temperature gradient and practically frictionless, then the process is irreversible. Then the heat and work transfers may be difficult to calculate, and irreversible thermodynamics is called for. Nevertheless, the first law still holds and provides a check on the measurements and calculations of the work done irreversibly on the system, $W_{A\to B}^{\text{path } P_1,\text{irreversible}}$, and the heat transferred irreversibly to the system, $Q_{A\to B}^{\text{path } P_1,\text{irreversible}}$, which belong to the same particular process defined by its particular irreversible path, P_1, through the space of thermodynamic states.

$$-W_{A\to B}^{\text{path } P_1,\text{irreversible}} + Q_{A\to B}^{\text{path } P_1,\text{irreversible}} = \Delta U$$

This means that the internal energy U is a function of state and that the internal energy change ΔU between two states is a function only of the two states.

Overview of the Weight of Evidence for the Law

The first law of thermodynamics is so general that its predictions cannot all be directly tested. In many properly conducted experiments it has been precisely supported, and never violated. Indeed, within its scope of applicability, the law is so reliably established, that, nowadays, rather than experiment being considered as testing the accuracy of the law, it is more practical and realistic to think of the law as testing the accuracy of experiment. An experimental result that seems to violate the law may be assumed to be inaccurate or wrongly conceived, for example due to failure to account for an important physical factor. Thus, some may regard it as a principle more abstract than a law.

State Functional Formulation for Infinitesimal Processes

When the heat and work transfers in the equations above are infinitesimal in magnitude, they are often denoted by δ, rather than exact differentials denoted by d, as a reminder that heat and work do not describe the *state* of any system. The integral of an inexact differential depends upon the particular path taken through the space of thermodynamic parameters while the integral of an exact differential depends only upon the initial and final states. If the initial and final states are the same, then the integral of an inexact differential may or may not be zero, but the integral of an exact differential is always zero. The path taken by a thermodynamic system through a chemical or physical change is known as a thermodynamic process.

The first law for a closed homogeneous system may be stated in terms that include concepts that are established in the second law. The internal energy U may then be expressed as a function of the system's defining state variables S, entropy, and V, volume: $U = U(S, V)$. In these terms, T, the system's temperature, and P, its pressure, are partial derivatives of U with respect to S and V. These variables are important throughout thermodynamics, though not necessary for the statement of the first law. Rigorously, they are defined only when the system is in its own state of internal thermodynamic equilibrium. For some purposes, the concepts provide good approximations for scenarios sufficiently near to the system's internal thermodynamic equilibrium.

The first law requires that:

$$dU = \delta Q - \delta W \text{ (closed system, general process, quasi-static or irreversible).}$$

Then, for the fictive case of a reversible process, dU can be written in terms of exact differentials. One may imagine reversible changes, such that there is at each instant negligible departure from thermodynamic equilibrium within the system. This excludes isochoric work. Then, mechanical work is given by $\delta W = - P\,dV$ and the quantity of heat added can be expressed as $\delta Q = T\,dS$. For these conditions,

$$dU = TdS - PdV \text{ (closed system, reversible process).}$$

While this has been shown here for reversible changes, it is valid in general, as U can be considered as a thermodynamic state function of the defining state variables S and V:

$$dU = TdS - PdV \text{ (closed system, general process, quasi-static or irreversible).}$$

Above equation is known as the fundamental thermodynamic relation for a closed system in the energy representation, for which the defining state variables are S and V, with respect to which T and P are partial derivatives of U. It is only in the fictive reversible case, when isochoric work is excluded, that the work done and heat transferred are given by $-P\,dV$ and $T\,dS$.

In the case of a closed system in which the particles of the system are of different types and, because chemical reactions may occur, their respective numbers are not necessarily constant, the fundamental thermodynamic relation for dU becomes:

$$dU = TdS - PdV + \sum_i \mu_i dN_i.$$

where dN_i is the (small) increase in amount of type-i particles in the reaction, and μ_i is known as the chemical potential of the type-i particles in the system. If dN_i is expressed in mol then μ_i is expressed in J/mol. If the system has more external mechanical variables than just the volume that can change, the fundamental thermodynamic relation further generalizes to:

$$dU = TdS - \sum_i X_i dx_i + \sum_j \mu_j dN_j.$$

Here the X_i are the generalized forces corresponding to the external variables x_i. The parameters X_i are independent of the size of the system and are called intensive parameters and the x_i are proportional to the size and called extensive parameters.

For an open system, there can be transfers of particles as well as energy into or out of the system during a process. For this case, the first law of thermodynamics still holds, in the form that the internal energy is a function of state and the change of internal energy in a process is a function only of its initial and final states, as noted in the section below headed First law of thermodynamics for open systems.

A useful idea from mechanics is that the energy gained by a particle is equal to the force applied to the particle multiplied by the displacement of the particle while that force is applied. Now consider the first law without the heating term: $dU = -PdV$. The pressure P can be viewed as a force (and in fact has units of force per unit area) while dV is the displacement (with units of distance times area). We may say, with respect to this work term, that a pressure difference forces a transfer of volume, and that the product of the two (work) is the amount of energy transferred out of the system as a result of the process. If one were to make this term negative then this would be the work done on the system.

It is useful to view the TdS term in the same light: here the temperature is known as a "generalized" force (rather than an actual mechanical force) and the entropy is a generalized displacement.

Similarly, a difference in chemical potential between groups of particles in the system drives a chemical reaction that changes the numbers of particles, and the corresponding product is the amount of chemical potential energy transformed in process. For example, consider a system consisting of two phases: liquid water and water vapor. There is a generalized "force" of evaporation that drives water molecules out of the liquid. There is a generalized "force" of condensation that drives vapor molecules out of the vapor. Only when these two "forces" (or chemical potentials) are equal is there equilibrium, and the net rate of transfer zero.

The two thermodynamic parameters that form a generalized force-displacement pair are called "conjugate variables". The two most familiar pairs are, of course, pressure-volume, and temperature-entropy.

Spatially Inhomogeneous Systems

Classical thermodynamics is initially focused on closed homogeneous systems (e.g. Planck 1897/1903), which might be regarded as 'zero-dimensional' in the sense that they have no spatial variation. But it is desired to study also systems with distinct internal motion and spatial inhomogeneity. For such systems, the principle of conservation of energy is expressed in terms not only of internal energy as defined for homogeneous systems, but also in terms of kinetic energy and potential energies of parts of the inhomogeneous system with respect to each other and with respect to long-range external forces. How the total energy of a system is allocated between these three more specific kinds of energy varies according to the purposes of different writers; this is because these components of energy are to some extent mathematical artefacts rather than actually measured physical quantities. For any closed homogeneous component of an inhomogeneous closed system, if E denotes the total energy of that component system, one may write,

$$E = E^{kin} + E^{pot} + U$$

where E^{kin} and E^{pot} denote respectively the total kinetic energy and the total potential energy of the component closed homogeneous system, and U denotes its internal energy.

Potential energy can be exchanged with the surroundings of the system when the surroundings impose a force field, such as gravitational or electromagnetic, on the system.

A compound system consisting of two interacting closed homogeneous component subsystems has a potential energy of interaction E_{12}^{pot} between the subsystems. Thus, in an obvious notation, one may write,

$$E = E_1^{kin} + E_1^{pot} + U_1 + E_2^{kin} + E_2^{pot} + U_2 + E_{12}^{pot}$$

The quantity E_{12}^{pot} in general lacks an assignment to either subsystem in a way that is not arbitrary, and this stands in the way of a general non-arbitrary definition of transfer of energy as work. On occasions, authors make their various respective arbitrary assignments.

The distinction between internal and kinetic energy is hard to make in the presence of turbulent motion within the system, as friction gradually dissipates macroscopic kinetic energy of localised bulk flow into molecular random motion of molecules that is classified as internal energy. The rate of dissipation by friction of kinetic energy of localised bulk flow into internal energy, whether in turbulent or in streamlined flow, is an important quantity in non-equilibrium thermodynamics. This is a serious difficulty for attempts to define entropy for time-varying spatially inhomogeneous systems.

First law of Thermodynamics for Open Systems

For the first law of thermodynamics, there is no trivial passage of physical conception from the closed system view to an open system view. For closed systems, the concepts of an adiabatic enclosure and of an adiabatic wall are fundamental. Matter and internal energy cannot permeate or penetrate such a wall. For an open system, there is a wall that allows penetration by matter. In general, matter in diffusive motion carries with it some internal energy, and some microscopic potential energy changes accompany the motion. An open system is not adiabatically enclosed.

There are some cases in which a process for an open system can, for particular purposes, be considered as if it were for a closed system. In an open system, by definition hypothetically or potentially, matter can pass between the system and its surroundings. But when, in a particular case, the process of interest involves only hypothetical or potential but no actual passage of matter, the process can be considered as if it were for a closed system.

Internal Energy for an Open System

Since the revised and more rigorous definition of the internal energy of a closed system rests upon the possibility of processes by which adiabatic work takes the system from one state to another, this leaves a problem for the definition of internal energy for an open system, for which adiabatic work is not in general possible. According to Max Born, the transfer of matter and energy across an open connection "cannot be reduced to mechanics". In contrast to the case of closed systems, for open systems, in the presence of diffusion, there is no unconstrained and unconditional physical distinction between convective transfer of internal energy by bulk flow of matter, the transfer of internal energy without transfer of matter (usually called heat conduction and work transfer), and change of various potential energies. The older traditional way and the conceptually revised (Carathéodory) way agree that there is no physically unique definition of heat and work transfer processes between open systems.

In particular, between two otherwise isolated open systems an adiabatic wall is by definition impossible. This problem is solved by recourse to the principle of conservation of energy. This principle allows a composite isolated system to be derived from two other component non-interacting isolated systems, in such a way that the total energy of the composite isolated system is equal to the sum of the total energies of the two component isolated systems. Two previously isolated systems can be subjected to the thermodynamic operation of placement between them of a wall permeable to matter and energy, followed by a time for establishment of a new thermodynamic state of internal equilibrium in the new single unpartitioned system. The internal energies of the initial two systems and of the final new system, considered respectively as closed systems as above, can be measured. Then the law of conservation of energy requires that,

$$\Delta U_s + \Delta U_o = 0,$$

Where ΔU_s and ΔU_o denote the changes in internal energy of the system and of its surroundings respectively. This is a statement of the first law of thermodynamics for a transfer between two otherwise isolated open systems, that fits well with the conceptually revised and rigorous statement of the law stated above.

For the thermodynamic operation of adding two systems with internal energies U_1 and U_2, to produce a new system with internal energy U, one may write $U = U_1 + U_2$; the reference states for U, U_1 and U_2 should be specified accordingly, maintaining also that the internal energy of a system be proportional to its mass, so that the internal energies are extensive variables.

There is a sense in which this kind of additivity expresses a fundamental postulate that goes beyond the simplest ideas of classical closed system thermodynamics; the extensivity of some variables is not obvious, and needs explicit expression; indeed one author goes so far as to say that it could be recognized as a fourth law of thermodynamics, though this is not repeated by other authors.

Also of course

$$\Delta N_s + \Delta N_o = 0,$$

where ΔN_s and ΔN_o denote the changes in mole number of a component substance of the system and of its surroundings respectively. This is a statement of the law of conservation of mass.

Process of Transfer of Matter Between an Open System and its Surroundings

A system connected to its surroundings only through contact by a single permeable wall, but otherwise isolated, is an open system. If it is initially in a state of contact equilibrium with a surrounding subsystem, a thermodynamic process of transfer of matter can be made to occur between them if the surrounding subsystem is subjected to some thermodynamic operation, for example, removal of a partition between it and some further surrounding subsystem. The removal of the partition in the surroundings initiates a process of exchange between the system and its contiguous surrounding subsystem.

An example is evaporation. One may consider an open system consisting of a collection of liquid, enclosed except where it is allowed to evaporate into or to receive condensate from its vapor above it, which may be considered as its contiguous surrounding subsystem, and subject to control of its volume and temperature.

A thermodynamic process might be initiated by a thermodynamic operation in the surroundings, that mechanically increases in the controlled volume of the vapor. Some mechanical work will be done within the surroundings by the vapor, but also some of the parent liquid will evaporate and enter the vapor collection which is the contiguous surrounding subsystem. Some internal energy will accompany the vapor that leaves the system, but it will not make sense to try to uniquely identify part of that internal energy as heat and part of it as work. Consequently, the energy transfer that accompanies the transfer of matter between the system and its surrounding subsystem cannot be uniquely split into heat and work transfers to or from the open system. The component of total energy transfer that accompanies the transfer of vapor into the surrounding subsystem is customarily called 'latent heat of evaporation', but this use of the word heat is a quirk of customary historical language, not in strict compliance with the thermodynamic definition of transfer of energy as heat. In this example, kinetic energy of bulk flow and potential energy with respect to long-range external forces such as gravity are both considered to be zero. The first law of thermodynamics refers to the change of internal energy of the open system, between its initial and final states of internal equilibrium.

Open System with Multiple Contacts

An open system can be in contact equilibrium with several other systems at once.

This includes cases in which there is contact equilibrium between the system, and several subsystems in its surroundings, including separate connections with subsystems through walls that are permeable to the transfer of matter and internal energy as heat and allowing friction of passage of the transferred matter, but immovable, and separate connections through adiabatic walls with others, and separate connections through diathermic walls impermeable to matter with yet others. Because there are physically separate connections that are permeable to energy but impermeable to matter, between the system and its surroundings, energy transfers between them can occur with definite heat and work characters. Conceptually essential here is that the internal energy transferred with the transfer of matter is measured by a variable that is mathematically independent of the variables that measure heat and work.

With such independence of variables, the total increase of internal energy in the process is then determined as the sum of the internal energy transferred from the surroundings with the transfer of matter through the walls that are permeable to it, and of the internal energy transferred to the system as heat through the diathermic walls, and of the energy transferred to the system as work through the adiabatic walls, including the energy transferred to the system by long-range forces. These simultaneously transferred quantities of energy are defined by events in the surroundings of the system. Because the internal energy transferred with matter is not in general uniquely resolvable into heat and work components, the total energy transfer cannot in general be uniquely resolved into heat and work components. Under these conditions, the following formula can describe the process in terms of externally defined thermodynamic variables, as a statement of the first law of thermodynamics:

$$\Delta U_0 = Q - W - \sum_{i=1}^{m} \Delta U_i$$

(suitably defined surrounding subsystems, general process, quasi-static or irreversible),

Where ΔU_o denotes the change of internal energy of the system, and ΔU_i denotes the change of internal energy of the *i*th of the *m* surrounding subsystems that are in open contact with the system, due to transfer between the system and that *i*th surrounding subsystem, and *Q* denotes the internal energy transferred as heat from the heat reservoir of the surroundings to the system, and *W* denotes the energy transferred from the system to the surrounding subsystems that are in adiabatic connection with it. The case of a wall that is permeable to matter and can move so as to allow transfer of energy as work is not considered here.

Combination of First and Second Laws

If the system is described by the energetic fundamental equation, $U_o = U_o(S, V, N_j)$, and if the process can be described in the quasi-static formalism, in terms of the internal state variables of the system, then the process can also be described by a combination of the first and second laws of thermodynamics, by the formula,

$$dU_0 = TdS - PdV + \sum_{j=1}^{n} \mu_j dN_j$$

Where there are *n* chemical constituents of the system and permeably connected surrounding subsystems, and where *T*, *S*, *P*, *V*, N_j, and μ_j, are defined as above.

For a general natural process, there is no immediate term-wise correspondence between above given equations, because they describe the process in different conceptual frames.

Nevertheless, a conditional correspondence exists. There are three relevant kinds of wall here: purely diathermal, adiabatic, and permeable to matter. If two of those kinds of wall are sealed off, leaving only one that permits transfers of energy, as work, as heat, or with matter, then the remaining permitted terms correspond precisely. If two of the kinds of wall are left unsealed, then energy transfer can be shared between them, so that the two remaining permitted terms do not correspond precisely.

For the special fictive case of quasi-static transfers, there is a simple correspondence. For this, it is supposed that the system has multiple areas of contact with its surroundings. There are pistons that allow adiabatic work, purely diathermal walls, and open connections with surrounding subsystems of completely controllable chemical potential (or equivalent controls for charged species). Then, for a suitable fictive quasi-static transfer, one can write.

$$\delta Q = TdS \quad \text{and} \quad \delta W = PdV$$

(suitably defined surrounding subsystems, quasi-static transfers of energy).

For fictive quasi-static transfers for which the chemical potentials in the connected surrounding subsystems are suitably controlled, these can be put into above equation to yield,

$$dU_0 = \delta Q - \delta W + \sum_{j=1}^{n} \mu_j dN_j$$

(suitably defined surrounding subsystems, quasi-static transfers).

The reference does not actually write above equation, but what it does write is fully compatible with it. Another helpful account is given by Tschoegl.

There are several other accounts of this, in apparent mutual conflict.

Non-equilibrium Transfers

The transfer of energy between an open system and a single contiguous subsystem of its surroundings is considered also in non-equilibrium thermodynamics. The problem of definition arises also in this case. It may be allowed that the wall between the system and the subsystem is not only permeable to matter and to internal energy, but also may be movable so as to allow work to be done when the two systems have different pressures. In this case, the transfer of energy as heat is not defined.

Methods for study of non-equilibrium processes mostly deal with spatially continuous flow systems. In this case, the open connection between system and surroundings is usually taken to fully surround the system, so that there are no separate connections impermeable to matter but permeable to heat. Except for the special case mentioned above when there is no actual transfer of matter, which can be treated as if for a closed system, in strictly defined thermodynamic terms, it follows that transfer of energy as heat is not defined. In this sense, there is no such thing as 'heat flow' for a continuous-flow open system. Properly, for closed systems, one speaks of transfer of internal energy as heat, but in general, for open systems, one can speak safely only of transfer of internal energy. A factor here is that there are often cross-effects between distinct transfers, for example that transfer of one substance may cause transfer of another even when the latter has zero chemical potential gradient.

Usually transfer between a system and its surroundings applies to transfer of a state variable, and obeys a balance law, that the amount lost by the donor system is equal to the amount gained by the receptor system. Heat is not a state variable. For his 1947 definition of "heat transfer" for discrete open systems, the author Prigogine carefully explains at some length that his definition of it does not obey a balance law. He describes this as paradoxical.

The situation is clarified by Gyarmati, who shows that his definition of "heat transfer", for continuous-flow systems, really refers not specifically to heat, but rather to transfer of internal energy, as follows. He considers a conceptual small cell in a situation of continuous-flow as a system defined in the so-called Lagrangian way, moving with the local center of mass. The flow of matter across the boundary is zero when considered as a flow of total mass. Nevertheless, if the material constitution is of several chemically distinct components that can diffuse with respect to one another, the system is considered to be open, the diffusive flows of the components being defined with respect to the center of mass of the system, and balancing one another as to mass transfer. Still there can be a distinction between bulk flow of internal energy and diffusive flow of internal energy in this case, because the internal energy density does not have to be constant per unit mass of material, and allowing for non-conservation of internal energy because of local conversion of kinetic energy of bulk flow to internal energy by viscosity.

Gyarmati shows that his definition of "the heat flow vector" is strictly speaking a definition of flow of internal energy, not specifically of heat, and so it turns out that his use here of the word heat is contrary to the strict thermodynamic definition of heat, though it is more or less compatible with historical custom, that often enough did not clearly distinguish between heat and internal energy; he writes "that this relation must be considered to be the exact definition of the concept of heat flow, fairly loosely used in experimental physics and heat technics." Apparently in a different frame of thinking from that of the above-mentioned paradoxical usage in the earlier sections of the historic 1947 work by Prigogine, about discrete systems, this usage of Gyarmati is consistent with the later sections of the same 1947 work by Prigogine, about continuous-flow systems, which use the term "heat flux" in just this way. This usage is also followed by Glansdorff and Prigogine in their 1971 text about continuous-flow systems. They write: "Again the flow of internal energy may be split into a convection flow $\rho u\mathrm{v}$ and a conduction flow. This conduction flow is by definition the heat flow W. Therefore: $\mathrm{j}[U] = \rho u\mathrm{v} + \mathrm{W}$ where u denotes the [internal] energy per unit mass. [These authors actually use the symbols E and e to denote internal energy but their notation has been changed here to accord with the notation of this article. These authors actually use the symbol *U* to refer to total energy, including kinetic energy of bulk flow.]" This usage is followed also by other writers on non-equilibrium thermodynamics such as Lebon, Jou, and Casas-Vásquez, and de Groot and Mazur. This usage is described by Bailyn as stating the non-convective flow of internal energy, and is listed as his definition number 1, according to the first law of thermodynamics. This usage is also followed by workers in the kinetic theory of gases. This is not the *ad hoc* definition of "reduced heat flux" of Haase.

In the case of a flowing system of only one chemical constituent, in the Lagrangian representation, there is no distinction between bulk flow and diffusion of matter. Moreover, the flow of matter is zero into or out of the cell that moves with the local center of mass. In effect, in this description, one is dealing with a system effectively closed to the transfer of matter. But still one can validly talk of a distinction between bulk flow and diffusive flow of internal energy, the latter driven by a temperature gradient within the flowing material, and being defined with respect to the local center of mass of the bulk flow. In this case of a virtually closed system, because of the zero matter transfer, as noted above, one can safely distinguish between transfer of energy as work, and transfer of internal energy as heat.

Various Heat Effects

1. Effect of temperature.

The variation of heat content in pure compounds with temperature can be determined by:

$$H_T - H_{298} = aT + bT^2 + cT^{-1} + d$$

$H_T - H_{298}$ is increase in heat content in cal/mole as the substance is heated from 298K to T. In all calculations the reference temperature is 298K (25°C or 77°F) T is temperature in K and a, b, c and d are constants and depends on state or aggregation of the system. The quantity $H_T - H_{298}$ is called sensible heat,

Molal heat capacity at constant pressure is,

$$C_p = a + 2bT - CT^{-2}$$

$$C_p = \left(\frac{\partial H}{\partial t}\right)_p \therefore H_2 - H_1 = \int_{T_1}^{T_2} C_p dT$$

Also change in heat content can be determined by mean heat capacity C_m:

$$H_2 - H_1 = C_m(T_2 - T_1)$$

$$C_m = \frac{\int_{T_1}^{T_2} C_p dT}{T_2 - T_1}$$

We may write:

$$H_2^1 - H_1^1 = C_m(T_2 - T_1)$$

Equation may be used to calculate heat content for rapid calculations. More accurate would be to use above equations.

1. Changes in state of aggregation

When solid is heated to melting point additional heat must be supplied to melt it. At the melting print, this additional heat does not increase the temperature. The heat content for melting at constant pressure is called latent heat of fusion (ΔH_{fusion}) and this heat must be supplied at the melting point to transform solid into liquid. An equal quantity of heat is liberated during solidification so that:

$$(\Delta H_{solidification}) = -(\Delta H_{fusion})$$

Negative sign indicates liberation of heat. Similarly, heat effects accompanying evaporation and allotropic changes in solids are measured by latent heat of vaporization and heats of transformation.

Value of ΔH for changes in state of aggregation vary with temperature and pressure under which change is carried out. For example heat of vaporization of water at 1000C is 542cal/g, whereas it is 583 cal/g at 25°C.

Example

Calculate latent heat of fusion of Cu.

The melting point of Cu is 1357K.

For solid $Cu: H_T - H_{298} = 5.41T + 0.75 \times 10^{-3} T^2 - 1680$

$= 7042\ cal/mol$ at melting point.

For liquid $C_u H_T - H_{298} = 7.50T - 20$

$= 10158\ cal/mol$ atmelting point.

Therefore $\Delta H_{fusion} = 10158 - 7042 = 3116 cal/mol.$

Heat of Formation

The formation of chemical compound from its elements is associated with either absorption or liberation of heat. Thus the formation of $H_2o(1)$ form the elements at 298K $(25\ C)$ can be written as,

$$H_2(g) + \frac{1}{2}O_2(g) = H_2O(1)$$

$\Delta H^0_{298} = -68320 \text{ cal / mol}$ at 1 atm pressure

Heat of Reaction

ΔH for any process depends only on the initial and final states and not on the path. If a process is divided into several steps, and ΔH is determined for each step, the algebraic sum of the ΔH values of all steps will be equal to ΔH for the original process. Consider an example to calculate heat of reaction of the following reaction:

$$Fe_2O_3 + 3C = 2Fe + 3CO$$

This Reaction Comprises of the Following Reactions

$$C + O = CO; \Delta H^\circ = -29160 \frac{\text{cal}}{\text{mole}}$$

$$2Fe + 1.50_2 = Fe_2O_3; \Delta H^\circ = -198500 \frac{\text{cal}}{\text{mole}}$$

Heat of formation for $Fe_2O_3 + 3C = 2Fe + 3CO$ reaction at 298K is:

$$(\Delta H^\circ_f)_{298} = 3\Delta H^\circ_{co} - \Delta H^\circ_{Fe_2O_3}$$

It must be noted that heat of formation of element is zero by using the value of ΔH^0 we get $(\Delta H^\circ_f)_{298} = 111020 \text{cal / mol} Fe_2O_3$ (endothermic), that means $Fe_2O_3 + 3C = 2Fe + 3CO$ reaction is accompanied by absorption of heat.

Consider the Reaction

$$Fe_2O_3 + 3CO = 2Fe + 3CO_2 \quad (\Delta H^\circ_f)_{co_2} = -97200 \frac{\text{cal}}{\text{mol}}$$

$$(\Delta H^\circ_f)_{298} = 3 \times (\Delta H^\circ_f)_{co_2} - (\Delta H^\circ_f)_{Fe_2O_3} - 3(\Delta H^\circ_f)CO$$

$= -5620 \text{cal / mol} Fe_2O_3$. This is exothermic reaction.

The heat of reaction calculated above are at 298K $(25^\circ C)$ Heat of reaction depends on temperature and can be determined,

$$(\Delta H^\circ_f)_{T_2} = (\Delta H_f)_{298} + \Sigma(H_{T2} - H_{298})_{products} - \Sigma(H_{T2} - H_{T1})_{reactants}$$

Material Balance

The material and heat balance for an operating process is an essential and routine plant record, in much the same way as income and expenditure record is a part of operating a business.

The material balance shows the weights and analysis of input and output materials and the calculated inputs and outputs of each of the important metal elements and compounds.

This accounting also serves as a check on plant data in that the various totals of inputs and outputs should be equal.

The material balance is the starting point for other calculations like energy requirements and cost calculations. From the energy calculations, one can design the energy flow path and identify energy losses to optimize energy consumption.

Basics of Materials Balance

Law of conservation of Mass.

Mass of isolated systems remains constants irrespective of the changes occurring within the system. Consider an open system of constant volume & mass.

Mass is transferred in to and out of the system; mass will accumulate if transfer rates are unequal Material balance at unsteady state is,

$$\begin{bmatrix} Rate\ of\ Mass \\ in \end{bmatrix} = \begin{bmatrix} Rate\ of\ Mass \\ out \end{bmatrix} + \begin{bmatrix} Rate\ of\ accumlation \\ of\ mass \end{bmatrix}$$

$$\frac{dm}{dt} = \dot{m}_{in} - \dot{m}_{out}$$

Material balance equation is written for each component involved in the process;

$$\frac{dm_1}{dt} = (\dot{m}_{in})_1 - (\dot{m}_{out})_1$$

$$\frac{dm_2}{dt} = (\dot{m}_{in})_2 - (\dot{m}_{out})_2$$

$$\frac{dm_3}{dt} = (\dot{m}_{in})_3 - (\dot{m}_{out})_3$$

$$\frac{dm_i}{dt} = (\dot{m}_{in})_i - (\dot{m}_{out})_i$$

For a systems with multiple input and output streams, the material balance equation for Ith component:

$$\frac{dm_i}{dt} = (\dot{m}_{in})_i - (\dot{m}_{out})_i$$

For steady state:

$$\Sigma(\dot{m}_{in})_i - \Sigma(\dot{m}_{out})_i$$

For chemically reacting systems in addition to the law of conservation of mass, following two laws must also hold:

1. Law of definite proportions:

A given chemical compound always contains the same constitute elements in the same weight proportions. True for stoichiometric compounds.

2. Law of multiple proportions:

If two elements can form more than one compound, then the respective weights of one element that combines with a given weight of the other are in the ratio small whole numbers.

Basics of Energy Balance

Energy balance at steady state is:

Heat input=Heat output.

All sources of heat input should be considered for example:

- Heat of reaction at 298 K.
- Heat of fusion and evaporation.
- Sensible heat in reactants.

Many a time specific heat capacity is required to calculate sensible heat. Also heat of formation of compound would be required.

Similarly all different sources of heat output must also be considered. To mention few:

- Sensible heat in solid products.
- Sensible heat in gases exiting the reactor.

- Sensible heat in liquid products.
- Heat losses.

In all heat calculations a convenient reference temperature 298K is selected.

Metal Extraction Principles

Pyrometallurgical extraction of metals from natural oxide reserves involve the following steps:

- Removal of oxygen from the valuable mineral.
- Separation of metal from the gangue minerals.

For sulphide ore an additional step to convert sulphide into oxide is required. Several unit processes are employed to extract metal from ore by pyrometallurgical route. These are: roasting, matte smelting, reduction smelting, converting and refining. Different unit processes are combined to extract metal for example.

Production of metal	Combination of unit processes
Copper	Roasting +matte smelting +converting +refining Flash smelting +converting +refining.
Zinc & lead	Roasting +reduction smelting + refining
Nickel	Roasting +matte smelting + reduction smelting +refining.
Pig iron	Reduction smelting
Steel	Reduction smelting +refining.
Hydrometallurgical extraction is suitable for lean ores .here leaching followed by separation of metal from solution are the principal unit processes. Some examples of Hydrometallurgical extraction are	
Copper:	Roasting +leaching +cementation+ electro wining
Zinc:	Roasting +leaching +cementation+ electrolysis
Aluminum:	Leaching + electrolysis
Magnesium:	Leaching + electrolysis
Titanium:	Reduction smelting +leaching +chlorination +reduction by Mg

Mineral Beneficiation

Mineral beneficiation is the first step in extraction of metal from natural resources. With the depletion of high grade metal ores it is important to increase the metal grade of an ore by physical methods; which are termed mineral beneficiation. The objectives of mineral beneficiation are:

- To increase the metal grade of ore.
- To reduce the amount of gangue minerals so that lower volume of slag forms in pyromettallurgical extraction of metals. Slag contains mostly gangue minerals.
- To decrease the thermal energy required to separate liquid metal from gangue minerals.
- To decrease the aqueous solution requirement in hydrometturgical extraction of metals.

In this chapter mineral beneficiation science and technology are briefly reviewed so that readers can apply materials balance principles.

What Constitutes Mineral Beneficiation?

Ore is an aggregate of minerals and contains valuable and gangue minerals. The mineral beneficiation involves separations of gangue minerals from ore and is done in the following two stages:

- Liberation of valuable mineral by size reduction technologies. In most ores the valuable minerals is distributed in the matrix of ore.
- Concentration technologies to separate the gangue minerals and to achieve increase in the content of the valuable mineral to increase the metal grade.

Sizes Reduction Technologies

Size reduction or communication is an important step and may be used:

- To produce particles of required sizes and shapes.
- To liberate valuable mineral so that it can be concentrated.
- To increase the surface area available for chemical reaction.

It is often said that the efficiency of energy utilization during fragmentation of solid particles is only about 1% with respect to the new surface created. Energy consumption represents major cost in the mineral processing operation.

Crushing and grinding are size reduction methods. Crushing is applied to subsequent size reduction down to about 25mm. In grinding finer size is produced. Grinding or milling is an important size reduction method. In grinding force is applied by a medium which could either balls or rods .Both dry and wet grinding is done Wet grinding has the following characteristics.

- It requires less power.
- It does not need dust central equipment.
- Wet grinding uses more steel grinding media to mill the material/per ton of product, as a result there occurs increase in erosion of the lining material.
- Water is required for wet grinding.

Material balance is important to determine:

- Amount of water in a milling circuit.
- % solid in slurry (slurry is a mixture of solid in water).

Both information are needed to determine the pump capacity to transport slurry. In wet milling water/solid ratio is important to control the viscosity of slurry. Too dilute slurry will lead to excessive wear of the medium .Too high a solid concentration results in cushioning of the medium. Percent solid in slurry can be determined by,

$$\%\text{solid in slurry} = \frac{100\rho_s(\rho_m - 100)}{\rho_m(\rho_s - 1000)}$$

ρ_s, ρ_m is density of solid and slurry respectively. For example if ρ_s, is $3000\ kg/m^3$ $\rho_m = 1500 kg/m^3$ and then slurry contains 33.3% solid according to above equation.

Wet milling gives the following advantages:

- Less power requirement.
- Required pollution as compared with dry milling.

After wet milling the milled product is classified by a hydro cyclone in to undersize (also termed underflow) and oversize (overflow) . The overflow is taken to the plant for concentration operation, for example flotation The undersize is recirculated after milling in a ball mill for further separation.

Concentration Technologies: Basics

The objectives of concentration technologies is to separate the valuable mineral from the gangue minerals .In all concentration methods feed is divided in three streams, namely concentrate, middling and tailings. Middlings are recycled within the plant and as such the plant output is two products, namely concentrate and tailings. Tailings are disposed whereas concentrate is sent to metal extraction.

Recovery and Grade

Recovery of the mineral in the concentrate and metal grade of the concentrate are important. Recovery is defined as,

$$\text{Re cov ery} = 100 \times \frac{\text{Amount of valuable min eral in concentrate}}{\text{amount of feed}}$$

Grade of the concentrate can be defined either mineral grade or metal grade. Since concentrate is employed for metal extraction, metal grade is important (Note that concentrate contains mineral but not metal).

$$\text{Metal grade} = \frac{\text{Amount of metal in concentrate}}{\text{amount of concentrate}} \times 100$$

Metal grade means grade of valuable metal of the mineral in the concentrate. For example in the concentrate of chalcopyrite the grade of Cu is important. Similarly in the concentrate of galena, the grade of Pb is important. It must be clearly understood that ore does not contain metal. Metal grade is used to give an idea about the removal of gangue minerals and removal of oxygen or sulphur. For example mineral grade of pure Fe_2O_3 is unity but metal grade (or iron grade) of pure mineral Fe_2O_3 is 70% which means 30% oxygen has to be removed to get iron.

Recovery of a mineral in the concentrate can be 100% if all the feed is diverted in to concentrate. But metal grade will be very low. The maximum metal grade of the concentrate can be that of corresponding pure mineral, for example Cu grade in pure $CuFeS_2$ is 34.1%, lead grade in pure Pbs is 86.6% Zn grade in pure ZnS is 67%. Consider 1000 kg feed of chalcopyrite which produces 1000 kg concentrate. The concentrate contains $500kg\ CuFeS_2 200kg\ Fe_2O_3, 200kg\ SiO_2$ and 10 kg Al_2O_3 .The analysis of feed is same as that of concentrate.

Now the recovery $CuFeS_2$ in concentrate is 100%. But Cu grade in the,

$$\text{concentrate} = \frac{500 \times 0.347}{1000} \times 100 = 17\%$$

Separation Efficiency

The feed of any concentration method is the milled product. The milled product is a mixture of particles of different Sizes, shapes and with different proportion of valuable mineral and gangue minerals. We have to separate particles containing valuable minerals for the maximum recovery. At the sametime metal grade should also be maximum of the concentrate.

In the following the valuable mineral in the concentrate is defined as metallic value of the valuable mineral.

Thus a concentrate is composed of metal in the valuable mineral +gangue. Let m is the metal grade of pure mineral and c is the metal grade of the concentrate.

$$\text{Gangue in concentrate} = (m-c)$$

$$\text{Gangue in feed} = (m-f)$$

f is the metal grade of the feed .If m_F is mass of feed and m_c is mass of concentrate

Recovery of gangue in concentrate $(R_g) = \frac{M_c(m-c)}{M_F(m-t)} \times 100$

Recovery of gangue in concentrate $(R_M) = \frac{M_c \times c}{M_F \times f} \times 100$

(SE) Separation efficiency $(R_{M-}R_g)100 = 100\frac{M_c}{M_f}\left[\frac{m(c-f)}{f(m-t)}\right]$

Illustration on Separation Efficiency

Concentrate	Tin Grade	Recovery
1	63%	62%
2	42%	72%
3	21%	78%

Tin grade in feed is 1% and tin grade of pure mineral is 78.76% calculate separation efficiency (SE):

$$SE = 100\frac{Mc}{MF}\left[\frac{m(c-f)}{f(m-f)}\right]$$

$$\frac{Mc}{MF} = \frac{f\,R_M}{c \times 100}$$

$$SE = \frac{f\,R_M}{c}\left[\frac{m(c-f)}{f(m-f)}\right]$$

Concentrate	Separation efficiency
1	61.8%
2	71.18%
3	75.24%

Note separation efficiency in concentrate 3 is high but tin grade is very low as compared to concentrate 1. But tin grade of concentrate is very high.

Concentration Methods

The most important processes are:

1. Gravity concentration.
2. Flotation.
3. Magnetic and electrostatic separation.

Gravity separation separates the minerals according to their different densities. It is used for the concentration of very heavy or very light minerals within a wide range of grain sizes.

In heavy media separation the density of pulp is intermediate between that of valuable mineral and gangue minerals. In that case light minerals float on top and the heavy minerals sink to the bottom of the pulp independent of particle size.

Other methods of gravity concentration utilize a combination of gravitational, inertial, frictional and viscous effects. Commonly used methods are jigging, washing tables, spirals etc.

Separation by flotation is based on the ability or lack of ability of different surfaces to be wetted by water. Hydrophobic minerals will cling to the air bubbles and rise with them, whereas the hydrophilic minerals will sink. Reagents like frothers, collectors, activators, depressors and conditioners are added to make the separation. Floatation has found its greatest application in the concentration of the sulphide minerals.

Electro statics and magnetic separation is based on differences in electrical conductivity of the mineral and magnetic properties the minerals respectively.

Gravity Separation

Gravity separation is an industrial method of separating two components, either a suspension, or dry granular mixture where separating the components with gravity is sufficiently practical: i.e. the components of the mixture have different specific weight. All of the gravitational methods are common in the sense that they all use gravity as the dominant force. The working principle is to lift the material by vacuum over an inclined vibrating screen covered deck. This results in the material being suspended in air while the heavier impurities are left behind on the screen and are discharged from the stone outlet. Gravity separation is used in a wide variety of industries, and can be most simply differentiated by the characteristics of the mixture to be separated - principally that of 'wet' i.e. - a suspension versus 'dry' -a mixture of granular product. Often other methods are applied to make the separation faster and more efficient, such as flocculation, coagulation and suction. The most notable advantages of the gravitational methods are their cost effectiveness and in some cases excellent reduction. Gravity separation is an attractive unit operation as it generally has low capital and operating costs, uses few if any chemicals that might cause environmental concerns and the recent development of new equipment enhances the range of separations possible.

Examples of Application

Agriculture: Gravity Separation tables are used for the removal of impurities, admixture, insect damage and immature kernels from the following examples: Wheat, Barley, Oilseed Rape, Peas, Beans, Cocoa Beans, Linseed. They can be used to separate and standardize coffee beans, cocoa beans, peanuts, corn, peas, rice, wheat, sesame and other food grains.

The Gravity Separator separates products of same size but with difference in specific weight. It has a vibrating rectangular deck, which makes it easy for the product to travel a longer distance, ensuring improved quality of the end product. The pressurized air in the deck enables the material to split according to its specific weight. As a result, the heavier particles travel to the higher level while the lighter particles travel to the lower level of the deck. It comes with easily adjustable air fans to control the volume of air distribution at different areas of the vibrating deck to meet the air supply needs of the deck. The table inclination, speed of eccentric motion and the feed rate can be precisely adjusted to achieve smooth operation of the machine.

Recycling: Gravity Separators are used to remove viable or valuable components from the recycling mixture i.e.: metal from plastic, rubber from plastic, different grades of plastic and these valuable materials are further recycled and reutilized.

Preferential Flotation

Heavy liquids such as hhh tetrabromoethane can be used to separate ores from supporting rocks by preferential flotation. The rocks are crushed, and while sand, limestone, dolomite, and other types of rock material will float on TBE, ores such as sphalerite, galena and pyrite will sink.

Clarification/Thickening

Clarification is a name for the method of separating fluid from solid particles. Often clarification is used along with flocculation to make the solid particles sink faster to the bottom of the clarification pool while fluid is obtained from the surface which is free of solid particles.

Thickening is the same as clarification except reverse. Solids that sink to the bottom are obtained and fluid is rejected from the surface.

The difference of these methods could be demonstrated with the methods used in waste water processing: in the clarification phase, sludge sinks to the bottom of the pool and clear water flows over the clear water grooves and continues its journey. The obtained sludge is then pumped into the thickeners, where sludge thickens farther and is then obtained to be pumped into digestion to be prepared into fertilizer.

Sinking Chamber

When clearing gases, an often used and mostly working method for clearing large particles is to blow it into a large chamber where the gas's velocity reduces and the solid particles start sinking to the bottom. This method is used mostly because of its cheap cost.

Froth Flotation

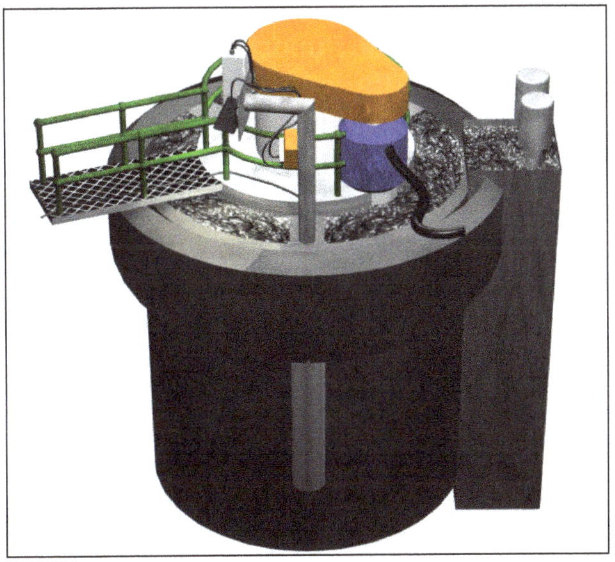

Diagram of a cylindrical froth flotation cell with camera and light used in image analysis of the froth surface.

Froth flotation is a process for selectively separating hydrophobic materials from hydrophilic. This is used in mineral processing, paper recycling and waste-water treatment industries. Historically this was first used in the mining industry, where it was one of the great enabling technologies of the 20th century. It has been described as "the single most important operation used for the recovery and upgrading of sulfide ores". The development of froth flotation has improved the recovery of valuable minerals, such as copper- and lead-bearing minerals. Along with mechanized mining, it has allowed the economic recovery of valuable metals from much lower grade ore than previously.

History

Initially, naturally occurring chemicals such as fatty acids and oils were used as flotation reagents in a large quantity to increase the hydrophobicity of the valuable minerals. Since then, the process has been adapted and applied to a wide variety of materials to be separated, and additional collector agents, including surfactants and synthetic compounds have been adopted for various applications.

William Haynes patented a process in 1869 for separating sulfide and gangue minerals using oil and called it *bulk-oil flotation*. In 1885 Carrie Everson expanded upon this and patented a process calling for oil[s] but also an acid or a salt.

The first successful commercial flotation process for mineral sulphides was invented by Frank Elmore who worked on the development with his brother, Stanley. The Glasdir copper mine at Llanelltyd, near Dolgellau in North Wales was bought in 1896 by the Elmore brothers in conjunction with their father, William. In 1897, the Elmore brothers installed the world's first industrial size commercial flotation process for mineral beneficiation at the Glasdir mine. The process was not froth flotation but used oil to agglomerate (make balls of) pulverised sulphides and buoy them to the surface, and was patented in 1898 with a description of the process published in the *Engineering and Mining Journal*. By this time they had recognized the importance of air bubbles in assisting the oil to carry away the mineral particles. The Elmores had formed a company known as the Ore Concentration Syndicate Ltd to promote the commercial use of the process worldwide. However developments elsewhere, particularly in Australia by Minerals Separation Ltd., led to decades of hard fought legal battles and litigations which, ultimately, were lost as the process was superseded by more advanced techniques. Charles Butters, beginning about 1899, and working with both the Elmores and Minerals Separation's representative E.H. Nutter developed what was known to contemporaries as the "Butters Process". The flotation process was independently invented in the early 1900s in Australia by Charles Vincent Potter and around the same time by Guillaume Daniel Delprat. This process (developed circa 1902) did not use oil, but relied upon flotation by the generation of gas formed by the introduction of acid into the pulp. In 1902, Froment combined oil and gaseous flotation using a modification of the Potter-Delprat process.

Another process was developed in 1902 by Cattermole, who emulsified the pulp with a small quantity of oil, subjected it to violent agitation, then slow stirring which coagulated the target minerals into nodules which were separated from the pulp by gravity. This was the basis of the Minerals Separation Ltd. process. By 1904, the MacQuisten process (a surface tension based method) was developed but this would not work when slimes were present. In 1912, James M. Hyde modified the Minerals Separation Process and installed it in the Butte and Superior Mill in Basin, Montana, the first such installation in the USA.

John M. Callow, of General Engineering of Salt Lake City, had followed flotation from technical papers and the introduction in both the Butte and Superior Mill, and at Inspiration Copper in Arizona and determined that mechanical agitation was a drawback to the existing technology. Introducing a porous brick with compressed air, and a mechanical stirring mechanism, Callow applied for a patent in 1914. This method, known as Pneumatic Flotation, was recognized to revolutionize the process of flotation concentration. The American Institute of Mining Engineers presented Callow the James Douglas Gold Medal in 1926 for his contributions to the field of flotation.

In the 1960s the froth flotation technique was adapted for deinking recycled paper.

Industries

Mining

Froth flotation to separate plastics, Argonne National Laboratory.

Froth flotation cells to concentrate copper and nickel sulfide minerals, Falconbridge, Ontario.

Froth flotation is a process for separating minerals from gangue by taking advantage of differences in their hydrophobicity. Hydrophobicity differences between valuable minerals and waste gangue are increased through the use of surfactants and wetting agents. The selective separation of the minerals makes processing complex (that is, mixed) ores economically feasible. The flotation process is used for the separation of a large range of sulfides, carbonates and oxides prior to further refinement. Phosphates and coal are also upgraded (purified) by flotation technology.

Prior to 1907, nearly all the copper mined in the US came from underground vein deposits, averaging 2.5 percent copper. By 1991, the average grade of copper ore mined in the US had fallen to only 0.6 percent.

Waste Water Treatment

The flotation process is also widely used in industrial waste water treatment plants, where it removes fats, oil, grease and suspended solids from waste water. These units are called dissolved air flotation (DAF) units. In particular, dissolved air flotation units are used in removing oil from the wastewater effluents of oil refineries, petrochemical and chemical plants, natural gas processing plants and similar industrial facilities.

Paper Recycling

Froth flotation is one of the processes used to recover recycled paper. In the paper industry this step is called deinking or just flotation. The target is to release and remove the hydrophobic contaminants from the recycled paper. The contaminants are mostly printing ink and stickies. Normally the setup is a two-stage system with 3,4 or 5 flotation cells in series.

Principle of Operation

Before froth flotation can work, the ore to be treated is reduced to fine particles by crushing and grinding (a process known as comminution) so that the various minerals exist as physically separate grains. This process is known as *liberation*. The particle sizes are typically less than 0.1 mm (100 µm), but sometimes sizes smaller than 7–10 µm are required. There is a tendency for the liberation size of the minerals to decrease over time as the ore bodies with coarse mineral grains that can be separated at larger sizes are depleted and replaced by ore bodies that were formerly considered too difficult.

In the mining industry, the plants where flotation is undertaken to concentrate ore are generally known as *concentrators* or *mills*.

For froth flotation, the ground ore is mixed with water to form a slurry and the desired mineral is rendered hydrophobic by the addition of a surfactant or *collector* chemical (although some mineral surfaces are naturally hydrophobic, requiring little or no addition of collector). The particular chemical depends on the nature of the mineral to be recovered and, perhaps, the natures of those that are not wanted. As an example, sodium ethyl xanthate may be added as a collector in the selective flotation of galena (lead sulfide) to separate it from sphalerite (zinc sulfide). This slurry (more properly called the *pulp*) of hydrophobic particles and hydrophilic particles is then introduced to tanks known as *flotation cells* that are aerated to produce bubbles. The hydrophobic particles attach to the air bubbles, which rise to the surface, forming a froth. The froth is removed from the cell, producing a concentrate ("con") of the target mineral.

Frothing agents, known as *frothers*, may be introduced to the pulp to promote the formation of a stable froth on top of the flotation cell.

The minerals that do not float into the froth are referred to as the *flotation tailings* or *flotation tails*. These tailings may also be subjected to further stages of flotation to recover the valuable particles that did not float the first time. This is known as *scavenging*. The final tailings after scavenging are normally pumped for disposal as mine fill or to tailings disposal facilities for long-term storage.

Froth flotation efficiency is determined by a series of probabilities: those of particle–bubble contact, particle–bubble attachment, transport between the pulp and the froth, and froth collection into the product launder. In a conventional mechanically-agitated cell, the void fraction (i.e. volume occupied by air bubbles) is low (5 to 10 percent) and the bubble size is usually greater than 1 mm. This results in a relatively low interfacial area and a low probability of particle–bubble contact. Consequently, several cells in series are required to increase the particle residence time, thus increasing the probability of particle–bubble contact.

Flotation is normally undertaken in several stages to maximize the recovery of the target mineral or minerals and the concentration of those minerals in the concentrate, while minimizing the energy input.

Flotation Stages

Roughing

The first stage is called *roughing*, which produces a *rougher concentrate*. The objective is to remove the maximum amount of the valuable mineral at as coarse a particle size as practical. The finer an ore is ground, the greater the energy that is required, so it makes sense to fine grind only those particles that need fine grinding. Complete liberation is not required for rougher flotation, only sufficient liberation to release enough gangue from the valuable mineral to get a high recovery.

The primary objective of roughing is to recover as much of the valuable minerals as possible, with less emphasis on the quality of the concentrate produced.

In some concentrators, there may be a *preflotation* step that precedes roughing. This is done when there are some undesirable materials, such as organic carbon, that readily float. They are removed first to avoid them floating during roughing (and thus contaminating the rougher concentrate).

Cleaning

The rougher concentrate is normally subjected to further stages of flotation to reject more of the undesirable minerals that also reported to the froth, in a process known as *cleaning*. The product of cleaning is known as the *cleaner concentrate* or the *final concentrate*.

The objective of cleaning is to produce as high a concentrate grade as possible.

The rougher concentrate is often subject to further grinding (usually called *regrinding*) to get more complete liberation of the valuable minerals. Because it is a smaller mass than that of the original ore, less energy is needed than would be necessary if the whole ore were reground. Regrinding is often undertaken in specialized *regrind mills*, such as the IsaMill™, designed to further reduce the energy consumed during regrinding to finer sizes.

Scavenging

The rougher flotation step is often followed by a *scavenger* flotation step that is applied to the rougher tailings. The objective is to recover any of the target minerals that were not recovered during the initial roughing stage. This might be achieved by changing the flotation conditions to make them more rigorous than the initial roughing, or there might be some secondary grinding to provide further liberation.

The concentrate from the rougher scavengers could be returned to the rougher feed for refloating or sent to special cleaner cells.

Similarly, the cleaning step may also be followed by a scavenging step performed on the cleaner tailings.

Science of Flotation

To be effective on a given ore slurry, the collectors are chosen based upon their selective wetting of the types of particles to be separated. A good collector will adsorb, physically or chemically, with one of the types of particles. This provides the thermodynamic requirement for the particles to bind to the surface of a bubble. The wetting activity of a surfactant on a particle can be quantified by measuring the contact angles that the liquid/bubble interface makes with it. Another important measure for attachment of bubbles to particles is induction time. The induction time is the time required for the particle and bubble to rupture the thin film separating the particle and bubble. This rupturing is achieved by the surface forces between the particle and bubble.

The mechanisms for the bubble-particle attachment is very complex and consists of three steps, collision, attachment and detachment. The collision is achieved by particles being within the collision tube of a bubble and this is affected by the velocity of the bubble and radius of the bubble. The collision tube corresponds to the region in which a particle will collide with the bubble, with the perimeter of the collision tube corresponding to the grazing trajectory.

The attachment of the particle to the bubble is controlled by the induction time of the particle and bubble. The particle and bubble need to bind and this occurs if the time in which the particle and bubble are in contact with each other is larger than the required induction time. This induction time is affected by the fluid viscosity, particle and bubble size and the forces between the particle and bubbles.

The detachment of a particle and bubble occurs when the force exerted by the surface tension is exceeded by shear forces and gravitational forces. These forces are complex and vary within the cell. High shear will be experienced close to the impeller of a mechanical flotation cell and mostly gravitational force in the collection and cleaning zone of a flotation column.

Significant issues of entrainment of fine particles occurs as these particles experience low collision efficiencies as well as sliming and degradation of the particle surfaces. Coarse particles show a low recovery of the valuable mineral due to the low liberation and high detachment efficiencies.

Theory

Selective Adhesion

Froth flotation depends on the selective adhesion of air bubbles to mineral surfaces in a mineral/water slurry. The air bubbles will attach to more hydrophobic particles. The attachment of the bubbles to the surface is determined by the interfacial energies between the solid, liquid, and gas phases. This is determined by the Young-Dupré Equation:

$$\gamma_{lv}\cos\theta = (\gamma_{sv} - \gamma_{sl})$$

where:

- γ_{lv} is the surface energy of the liquid/vapor interface,
- γ_{sv} is the surface energy of the solid/vapor interface,
- γ_{sl} is the surface energy of the solid/liquid interface,
- θ is the contact angle, the angle formed at the junction between vapor, solid, and liquid phases.

Minerals targeted for separation may be chemically surface-modified with collectors so that they are more hydrophobic. Collectors are a type of surfactant that increase the natural hydrophobicity of the surface, increasing the separability of the hydrophobic and hydrophilic particles. Collectors either chemically bond via chemisorption to the mineral or adsorb onto the surface via physisorption.

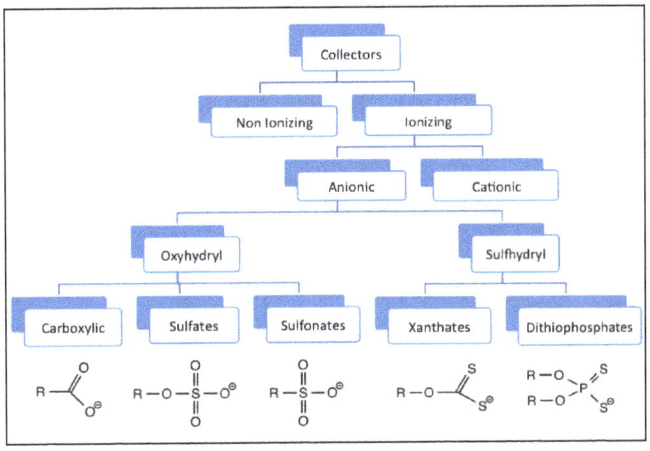

Different types of collectors, or surfactants, used in froth flotation.

IMFs and surface forces in Bubble-particle Interactions

Collision

The collision rates for fine particles (50 - 80 μm) can be accurately modeled, but there is no current theory that accurately models bubble-particle collision for particles as large as 300 μm, which are commonly used in flotation processes.

For fine particles, Stokes law underestimates collision probability while the potential equation based on Surface Charge overestimates collision probability so an intermediate equation is used.

It is important to know the collision rates in the system since this step precedes the adsorption where a three phase system is formed.

Adsorption (Attachment)

The effectiveness of a medium to adsorb to a particle is influenced by the relationship between the surfaces both materials. There are multiple factors that affect the efficiency of adsorption in chemical, thermodynamic, and physical domains. These factors can range from surface energy and polarity to the shape, size, and roughness of the particle. In froth flotation, adsorption is a strong consequence of surface energy, since the small particles have a high surface area to size ratio, resulting in higher energy surfaces to form attractions with adsorbates. The air bubbles must selectively adhere to the desired minerals to elevate them to the surface of the slurry while wetting the other minerals and leaving them in the aqueous slurry medium.

Particles that can be easily wetted by water are called hydrophilic, while particles that are not easily wetted by water are called hydrophobic. Hydrophobic particles have a tendency to form a separate phase in aqueous media. In froth flotation the effectiveness of an air bubble to adhere to a particle is based on how hydrophobic the particle is. Hydrophobic particles have an affinity to air bubbles, leading to adsorption. The bubble-particle combinations are elevated to the froth zone driven by buoyancy forces.

The attachment of the bubbles to the particles is determined by the interfacial energies of between the solid, liquid, and vapor phases, as modeled by the Young/Dupre Equation. The interfacial energies can be based on the natural structure of the materials, or the addition of chemical treatments can improve energy compatibility.

Collectors are the main additives used to improve particle surfaces. They function as surfactants to selectively isolate and aid adsorption between the particles of interest and bubbles rising through the slurry. Common collectors used in flotation are anionic sulfur ligands, which have a bifunctional structure with an ionic portion which shares attraction with metals, and a hydrophobic portion such as a long hydrocarbon tail. These collectors coat a particle's surface with a monolayer of non-polar substance to aid separation from the aqueous phase by decreasing the adsorbed particle solubility in water. The adsorbed ligands can form micelles around the particles and form small-particle colloids improving stability and phase separation further.

Desorption (Detachment)

The adsorption of particles to bubbles is essential to separating the minerals from the slurry, but the minerals must be purified from the additives used in separation, such as the collectors, frothers, and modifiers. The product of the cleaning, or desorption process, is known as the cleaner concentrate. The detachment of a particle and bubble requires adsorption bond cleavage driven by shear forces. Depending on the flotation cell type, shear forces are applied by a variety of mechanical systems. Among the most common are impellers and mixers. Some systems combine the functionalities of these components by placing them at key locations where they can take part in multiple froth flotation mechanisms. Cleaning cells also take advantage of gravitational forces to improve separation efficiency.

Performance Calculations

Relevant Equations

A common quantity used to describe the collection efficiency of a froth flotation process is *flotation recovery* (R). This quantity incorporates the probabilities of collision and attachment of particles to gas flotation bubbles.

$$R = \frac{N_c}{\left(\frac{\pi}{4}\right)\left(d_p + d_b\right)^2 Hc}$$

where:

- $N_c = PN_c^i$, which is the product of the probability of the particle being collected (P) and the number of possible particle collisions (N_c^i).
- d_p is particle diameter.
- d_b is bubble diameter.
- H is a specified height within the flotation which the recovery was calculated.
- c is the particle concentration.

The following, are several additional mathematical methods often used to evaluate the effectiveness of froth flotation processes. These equations are more simple than the calculation for *flotation recovery*, as they are based solely on the amounts of inputs and outputs of the processes.

For the following equations:

- F is the weight percent of feed.
- C is the weight percent concentrate.
- T is the weight percent of tailings.
- C, t, and f are the Metallurgical assays of the concentrate, tailings, and feed, respectively.

Ratio of feed weight to concentrate weight $\frac{F}{C}$ (unitless):

$$\frac{F}{C}=\frac{c-t}{f-t}$$

Percent of metal recovered (X_R) in wt%:

$$X_R=100\left(\frac{c}{f}\right)\left(\frac{f-t}{c-t}\right)$$

Percent of metal lost (X_L) in wt%:

$$X_L=100-X_R$$

Percent of weight recovered (X_W) in wt%:

$$X_W=100\left(\frac{C}{F}\right)=100\frac{f-t}{c-t}$$

Grade-recovery Curves

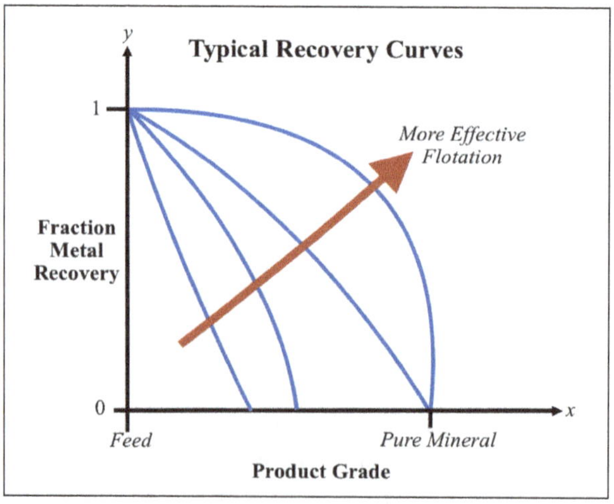

Example grade-recovery relationships seen in froth flotation. Shifts in the curves represent changes in flotation effectiveness.

Grade-recovery curves are useful tools in weighing the trade-off of producing a high grade of concentrate while maintaining as low of a recovery rate as possible, two important aspects of froth flotation. These curves are developed empirically based on the individual froth flotation process of a particular plant. As the curves are shifted in the positive x-direction (to the right) and the positive y-direction (upward) the performance of the froth flotation process is regarded as improving. A disadvantage to these curves is that they can only compare the grade-recovery relations of a specific feed grade and feed rate. If a company has a variance of feed grades and rates used (an extremely common occurrence) in their froth flotation process, grade-recovery curves for every pairing of feed grade and recovery rate would have to be constructed in order to provide meaningful information to the plant.

Flotation Equipment

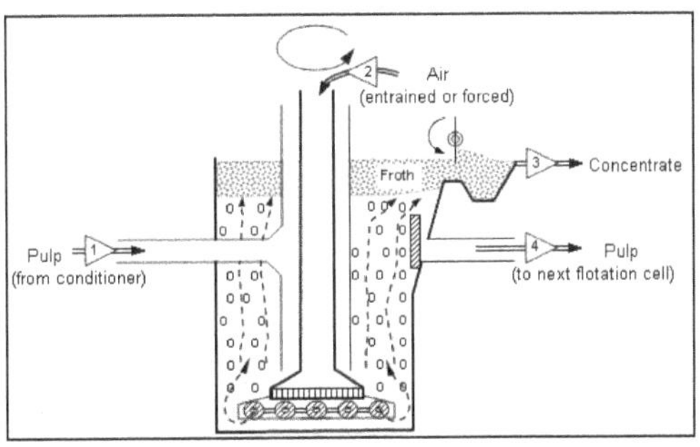

Diagram of froth flotation cell. Numbered triangles show direction of stream flow. A mixture of ore and water called pulp enters the cell from a conditioner, and flows to the bottom of the cell. Air or nitrogen is passed down a vertical impeller where shearing forces break the air stream into small bubbles. The mineral concentrate froth is collected from the top of the cell , while the pulp flows to another cell.

Flotation can be performed in rectangular or cylindrical mechanically agitated cells or tanks, flotation columns, Jameson Cells or deinking flotation machines. Classified by the method of air absorption manner, it is fair to state that two distinct groups of flotation equipments have arisen:pneumatic and mechanical machines. Generally pneumatic machines give a low-grade concentrate and little operating troubles.

Comparison of the sizes of flotation columns and Jameson Cells with similar capacities.

Mechanical cells use a large mixer and diffuser mechanism at the bottom of the mixing tank to introduce air and provide mixing action. Flotation columns use air spargers to introduce air at the bottom of a tall column while introducing slurry above. The countercurrent motion of the slurry flowing down and the air flowing up provides mixing action. Mechanical cells generally have a higher throughput rate, but produce material that is of lower quality, while flotation columns generally have a low throughput rate but produce higher quality material.

The Jameson cell uses neither impellers nor spargers, instead combining the slurry with air in a downcomer where high shear creates the turbulent conditions required for bubble particle contacting.

Mechanics of Flotation

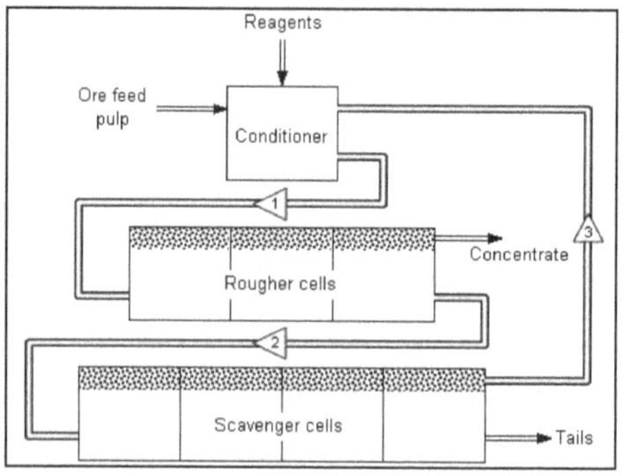

The following steps are followed, following grinding to liberate the mineral particles:

1. Reagent conditioning to achieve hydrophobic surface charges on the desired particles.
2. Collection and upward transport by bubbles in an intimate contact with air or nitrogen.
3. Formation of a stable froth on the surface of the flotation cell.
4. Separation of the mineral laden froth from the bath (flotation cell).

Simple flotation circuit for mineral concentration. Numbered triangles show direction of stream flow. Various flotation reagents are added to a mixture of ore and water (called pulp) in a conditioning tank. The flow rate and tank size are designed to give the minerals enough time to be activated. The conditioner pulp is fed to a bank of rougher cells which remove most of the desired minerals as a concentrate. The rougher pulp passes to a bank of scavenger cells where additional reagents may be added. The scavenger cell froth is usually returned to the rougher cells for additional treatment, but in some cases may be sent to special cleaner cells. The scavenger pulp is usually barren enough to be discarded as tails. More complex flotation circuits have several sets of cleaner and re-cleaner cells, and intermediate re-grinding of pulp or concentrate.

Chemicals of Flotation

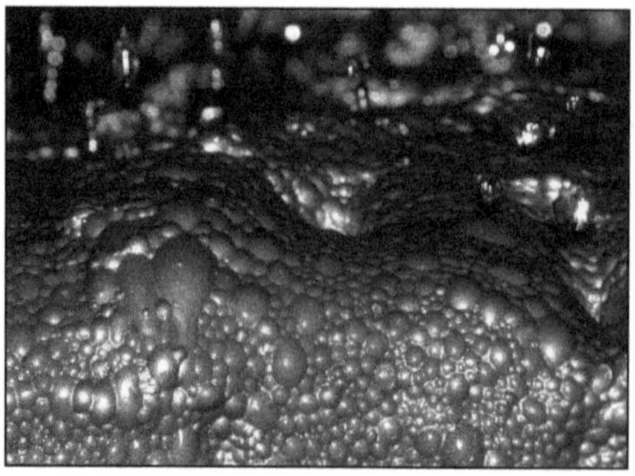

Copper sulfide foam in a froth-flotation cell

Collectors

For many ores (e.g. those of Cu, Mo, W, Ni), the collectors are anionic sulfur ligands. Particularly popular are xanthate salts, including potassium amyl xanthate (PAX), potassium isobutyl xanthate (PIBX), potassium ethyl xanthate (KEX), sodium isobutyl xanthate (SIBX), sodium isopropyl xanthate (SIPX), sodium ethyl xanthate (SEX). Other collectors include related sulfur-based ligands: dithiophosphates, dithiocarbamates. Still other classes of collectors include the thioureathiocarbanilide. Fatty acids have also been used.

For some minerals (e.g., sylvanite for KCl), fatty amines are used as collectors.

Frothers

A variety of compounds are added to stabilize the foams. These additives include pine oil, various alcohols (methyl isobutyl carbinol (MIBC)), polyglycols, xylenol (cresylic acid).

Modifiers

A variety of other compounds are added to optimize the separation process, these additives are called modifiers. Such species do not interact directly with the mineral, but modify the physical properties of the solution.

- pH modifiers include lime (CaO), Soda ash (Na_2CO_3), Caustic soda (NaOH), sulfuric and hydrochloric acid (H_2SO_4, HCl).
- Anionic modifiers include phosphates, silicates, and carbonates.
- Organic modifiers include the thickeners dextrin, starch, glue, and CMC.

Chemical Compounds for Deinking of Recycled Paper

- pH control: sodium silicate and sodium hydroxide.
- Calciumion source: hard water, lime or calcium chloride.
- Collector: fatty acid, fatty acid emulsion, fatty acid soap and/or organo-modified siloxane.

Specific ore Applications

Illustrative, the flotation process is used for purification of potassium chloride from sodium chloride and clay minerals. The crushed mineral is suspended in brine in the presence of fatty ammonium salts. Because the ammonium group and K^+ have very similar ionic radii (ca. 0.135, 0.143 nm respectively), the ammonium centers exchange for the surface potassium sites on the particles of KCl, but not on the NaCl particles.

The long alkyl chains then confer hydrophobicity, to the particles, which enable them to form foams.

Sulfide ores		
• Copper	• Copper-Molybdenum	• Lead-Zinc
• Lead-Zinc-Iron	• Copper-Lead-Zinc-Iron	• Gold-Silver
• Oxide Copper and Lead	• Nickel	• Nickel-Copper
Nonsulfide ores		
• Fluorite	• Tungsten	• Lithium
• Tantalum	• Tin	• Coal

References

- Nguyen, Anh V (12 June 1996). “On modelling of bubble–particle attachment probability in flotation”. International Journal of Mineral Processing. 53: 225–249. doi:10.1016/S0301-7516(97)00073-2
- What’s in a Name? Amount of Substance, Chemical Amount, and Stoichiometric Amount Carmen J. Giunta Journal of Chemical Education 2016 93 (4), 583-586 doi:10.1021/acs.jchemed.5b00690
- Tro, N. J. (2008). Chemistry. A Molecular Approach, Pearson/Prentice Hall, Upper Saddle River NJ, ISBN 0-13-100065-9, p. 246
- “CODATA Value: molar gas constant”. The NIST Reference on Constants, Units, and Uncertainty. US National Institute of Standards and Technology. June 2015. Retrieved 2015-09-25. 2014 CODATA recommended values

4

Metallurgical Processes

The extraction of metals from their respective ores and their further refinement can be undertaken through various methods. Bayer process is one such method through which Alumina is produced from bauxite. Bessemer process, Betts electrolyte process, patio process, Pidgeon process, Hoopes process and Dow process are some other metallurgical processes which are closely examined in this chapter.

Bayer Process

Alumina is the main components of bauxite, so bauxite is refined in order to produce Alumina. The Bayer process is the principal way for producing alumina by refining Bauxite. Bauxite other than with 30-60percentage of aluminium oxide contains mixture of silica, iron oxides and titanium dioxide. The Bayer Process is the most economic means of obtaining alumina from bauxite. Other processes for obtaining alumina from metal ores are also in use in some refineries, particularly in China and Russia, although these make up a relatively small percentage of global production. Prior to the Bayer process bauxite is crushed and ground in mills to fine particles (max. size0.06"/1.5mm). A hot solution of the recycled sodium hydroxide (caustic soda, NaOH) is then added to the ground ore forming a bauxite slurry, which is stored in holding tanks and then pumped to the further processing stage.

The Bayer process can beconsidered in three stages:

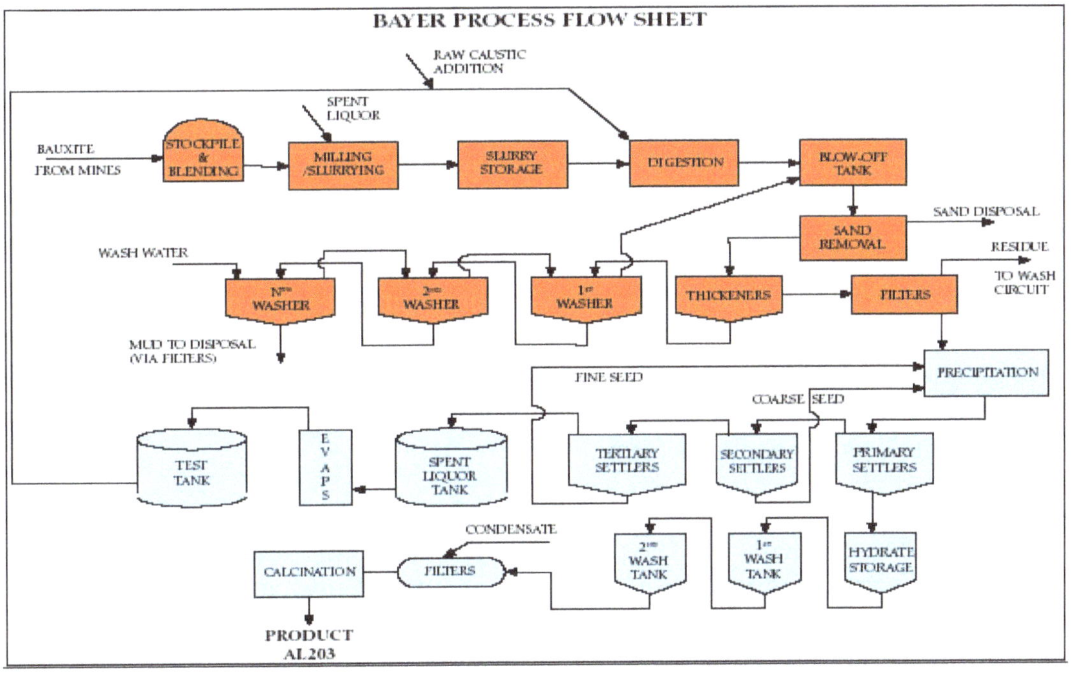

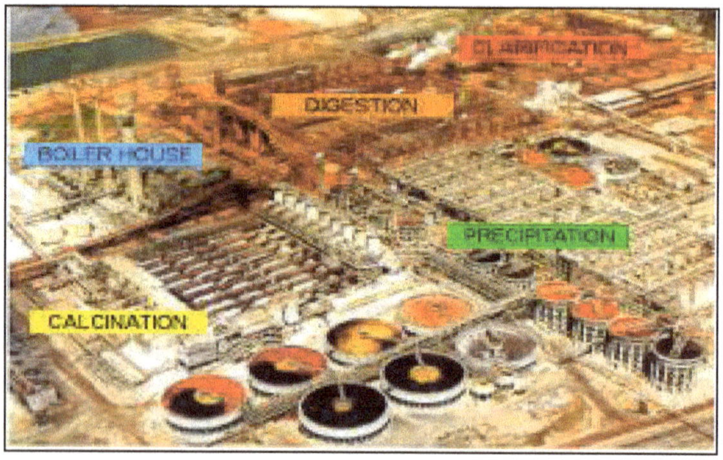

Figure: Aerial view of a Bayer processing plant.

Extraction

The aluminium-bearing minerals in bauxite - Gibbsite, Böhmite and Diaspore -are selectively extracted from the insoluble components (mostly oxides) by dissolving them in a solution of sodium hydroxide (caustic soda):

The first step in the process is the mixing of ground bauxite into a solution of sodium hydroxide. By applying steam and pressure in tanks containing the mixture, the bauxite slowly dissolves. The alumina released reacts with the sodium hydroxide to form sodium aluminate.

Gibbsite:

$$Al(OH)_3 + Na^+ + OH^- \rightarrow Al(OH)_4^- + Na^+$$

Böhmite and Diaspore:

$$AlO(OH) + Na^+ + OH^- + H_2O \rightarrow Al(OH)_4^- + Na^+$$

Depending on the quality of the ore it may be washed to beneficiate it prior to processing. The ore is crushed and milled to reduce the particle size and make the minerals more amenable to extraction. It is then combined with the process liquor and sent in a slurry form to a heated pressure digester.

Conditions within the digester (concentration, temperature and pressure) are set according to the properties of the bauxite ore. Ores with a high Gibbsite content can be processed at 140° C with about 110gl^{-1} of Na_2O. Processing of Böhmite on the other hand requires harsher conditions, between 200 and 240°C and 140-170 gl^{-1} of Na_2O. The pressure is not important for the process, as such but is defined by the steam pressure during the actual process conditions. At 240°C the pressure is approximately 35 atmospheres (atm).

Although higher temperatures are often theoretically advantageous there are several disadvantages including corrosion problems and the possibility of oxides other than alumina dissolving into the caustic liquor, leading to impurities problems in the final product, and scaling of pipes, pumps and tanks.

After digestion about 30% of the bauxite mass remains in suspension as a thin red mud slurry of silicates, and oxides of iron and titanium. The mud-laden liquor leaving the digestion vessel is flash-cooled to atmospheric boiling point by flowing through a series of flash vessels which operate at successively lower pressures.

The flash steam generated is used to preheat incoming caustic liquor in tubular heat exchangers located parallel to the flash tank line. Condensate from the heat exchangers is used for boiler feed water and washing waste mud.

After the extraction stage the insoluble bauxite residue must be separated from the Aluminium-containing liquor by a process known as settling. The mixture of solid impurities is called red mud, and presents a disposal problem. The liquor is purified as much as possible through filters before being transferred to the precipitators. Slaked lime is added to dilute caustic liquor in the washing process to remove carbonate (Na2CO3) which forms by reaction with compounds in bauxite and also from the atmosphere and which reduces the effectiveness of liquor to dissolve alumina. Lime regenerates caustic soda, allowing the insoluble calcium carbonate to be removed with the waste mud.

The insoluble red mud from the first settling stage is thickened and washed to recover the caustic soda, which is then recycled back into the main process.

The mud is washed with fresh water in counter-current washing trains to recover the soda and alumina content in the mud before being pumped to large disposal dams.

Precipitation

Crystalline aluminium trihydroxide (Gibbsite), conveniently named "hydrate", is then precipitated from the digestion liquor:

$$Al(OH)_4^- + Na^+ \rightarrow Al(OH)_3 + Na^+ + OH^-$$

This is basically the reverse of the extraction process, except that the product's nature is carefully controlled by plant conditions, including seeding or selective nucleation, precipitation temperature and cooling rate. The cooled pregnant liquor flows to rows of precipitation tanks which are seeded with crystalline trihydrate alumina, usually of an intermediate or fine particle size to promote crystal growth. Each precipitation tank is agitated, with a holding time of about three hours. During the 25-30 hours pass through precipitation, alumina of various crystal sizes is produced. The entry temperature and the temperature gradient across the row, seed rate and caustic concentration are control variables used to achieve the required particle size distribution in the product.

The finished mix of crystal sizes is settled from the liquor stream and separated into three size ranges in three stages "gravity" classification tanks. The primary classifiers collect the coarse fraction which becomes the product "hydrate". The intermediate and fine crystals from the secondary and tertiary classifiers are washed and returned to the precipitation tanks as coarse and fine seed. The process engineer decides on the ratio of coarse and fine seed to be added, based on screen analyses.

The slurry of coarse "hydrate" ($Al_2O_3.3H_2O$) from the primary thickeners is pumped to hydrate storage tanks and is filtered and washed on horizontal-table vacuum filters to remove and recycle process liquor, prior to calcination in fluid bed or rotary calciners. Heating the slurry before filtration, with waste steam, gives a cleaner, drier filter cake and leads to less soda loss. The heated cake entering the calciner also helps fuel efficiency.

Calcination

"Hydrate", is calcined to form alumina for the aluminium smelting process. In the calcination process water is driven off to form alumina, this takes place at 1050° C:

$$2Al(OH)_3 \rightarrow Al_2O_3 + 3H_2O$$

The calcination process must be carefully controlled since it dictates the properties of the final product. Sandy alumina particles are 90%+ 45 μm (microns) in size, this is the product sought after by most smelters.

A large amount of the alumina so produced is then subsequently smelted in the Hall-Héroult process, named after inventors Charles Hall and Paul-Louis Héroult, in order to produce aluminium.

Smelting

The purified Al2O3 is dissolved in molten cryolite, Na3AlF6, which has a melting point of 10120C and is an effective conductor of electric current. In the following schematic diagram of the electrolysis cell graphite rods are employed as anodes and are consumed in the electrolysis process. The cell electrolytic reaction is:

$$2Al_2O_3 + 3C \rightarrow 4Al(l) + 3CO_2(g)$$

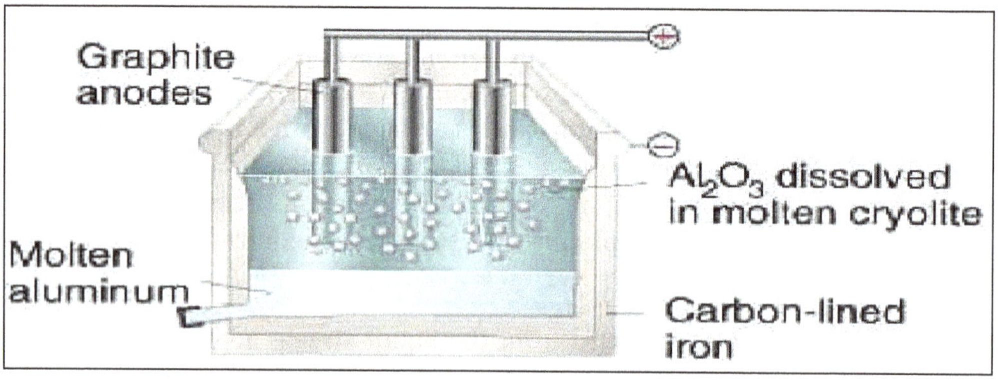

The production of one ton of aluminum requires about 65-70 GJ (18-20 MWh) and about half a ton of carbon.

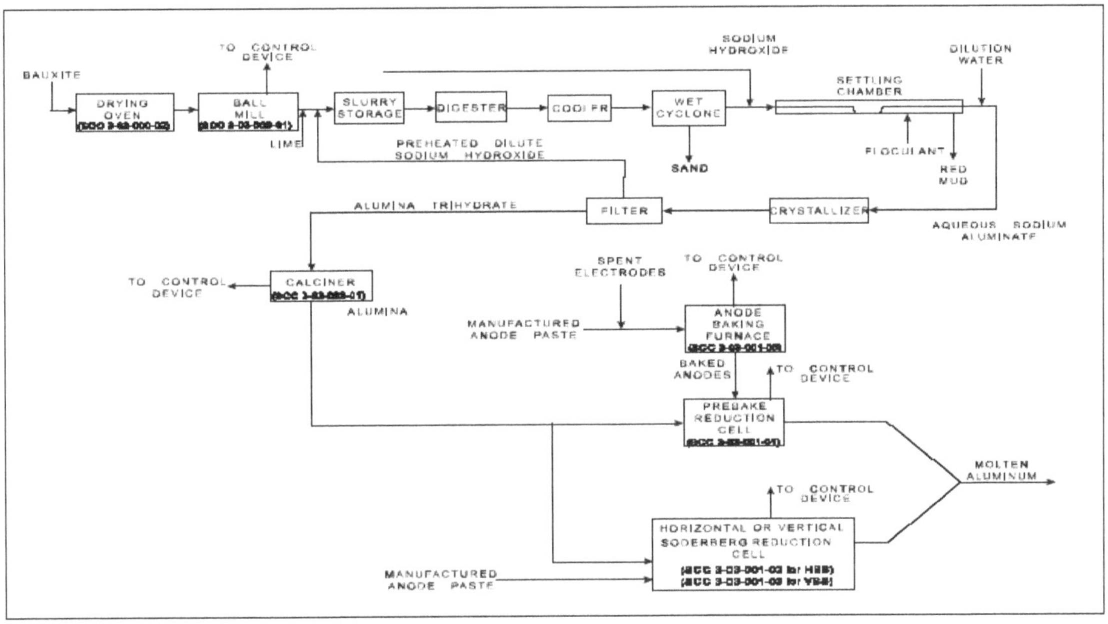

Figure: Schematic diagram of aluminium production.

Bessemer Process

The Bessemer process was the first inexpensive industrial process for the mass-production of steel from molten pig iron. The process is named after its inventor, Henry Bessemer, who took out a patent on the process in 1855. The key principle is removal of impurities from the iron by oxidation with air being blown through the molten iron. The oxidation also raises the temperature of the iron mass and keeps it molten.

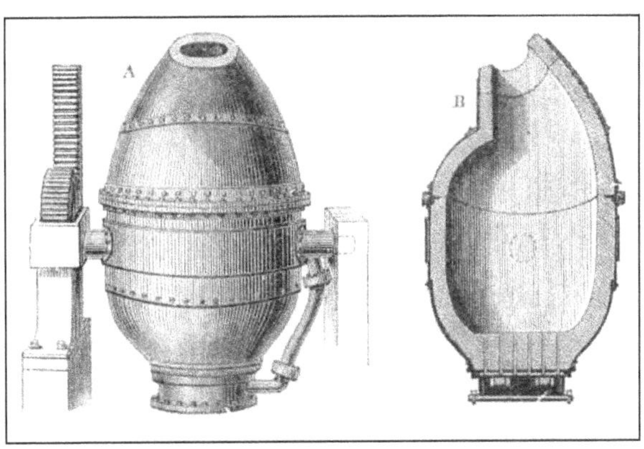

The process is carried on in a large ovoid steel container lined with clay or dolomite called the Bessemer converter. The capacity of a converter was from 8 to 30 tons of molten iron with a usual charge being around 15 tons. At the top of the converter is an opening, usually tilted to the side relative to the body of the vessel, through which the iron is introduced and the finished product removed. The bottom is perforated with a number of channels called tuyères through which air is forced into the converter. The converter is pivoted on trunnions so that it can be rotated to receive the charge, turned upright during conversion, and then rotated again for pouring out the molten steel at the end.

Oxidation

The oxidation process removes impurities such as silicon, manganese, and carbon as oxides. These oxides either escape as gas or form a solid slag. The refractory lining of the converter also plays a role in the conversion—the clay lining is used in the acid Bessemer, in which there is low phosphorus in the raw material. Dolomite is used when the phosphorus content is high in the basic Bessemer (limestone or magnesite linings are also sometimes used instead of dolomite). In order to give the steel the desired properties, other substances could be added to the molten steel when conversion was complete, such as spiegeleisen (an ironcarbon-manganese alloy).

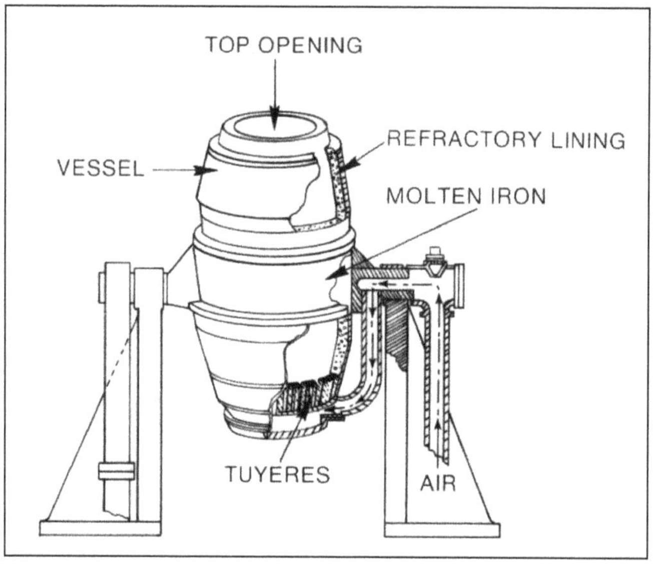

Managing the Process

When the required steel had been formed, it was poured out into ladles and then transferred into moulds and the lighter slag is left behind. The conversion process called the "blow" was completed in around twenty minutes. During this period the progress of the oxidation of the impurities was judged by the appearance of the flame issuing from the mouth of the converter.

After the blow, the liquid metal was recarburized to the desired point and other alloying materials are added, depending on the desired product.

Before the Bessemer process, Britain had no practical method of reducing the carbon content of pig iron. Steel was manufactured by the reverse process of adding carbon to carbon-free wrought iron, usually imported from Sweden. The manufacturing process, called cementation process consisted of heating bars of wrought iron together with charcoal for periods of up to a week in a long stone box. This produced blister steel. Up to 3 tons of expensive coke was burnt for each ton of steel produced. Such steel when rolled into bars was sold at £50 to £60 a long ton. The most difficult and work-intensive part of the process, however, was the production of wrought iron done in finery forges in Sweden.

This process was refined in the 1700s with the introduction of Benjamin Huntsman's crucible steel-making technique, which added an additional three hours firing time and required additional large quantities of coke. In making crucible steel the blister steel bars were broken into pieces and melted in small crucibles each containing 20 kg or so. This produced higher quality crucible steel but increased the cost. The Bessemer process reduced to about half an hour the time needed to make steel of this quality while requiring only the coke needed to melt the pig iron initially. The earliest Bessemer converters produced steel for £7 a long ton, although it initially sold for around £40 a ton.

Sir Henry Bessemer described the origin of his invention in his autobiography. at the time of the outbreak of the Crimean War many English industrialists and inventors became interested in military technology and Bessemer himself developed a method for grooving artillery projectiles so that they could spin without the use of rifling in the bore of the gun. He patented this method in 1854 and began developing it in conjunction with the government of France. After a successful day of testing of his method at the Polygon in France he had a conversation with Claude Etienne Minié who stated that a key barrier to the use of the larger, heavier spinning projectiles would be the strength of the gun and in particular "he [Minié] did not consider it safe in practice to fire a 30-lb. shot from a 12-pounder castiron gun. The real question, he said, was; Could any guns be made to stand such heavy projectiles?". This is what started Bessemer thinking about steel.

At the time steel was difficult and expensive to make and was consequently used in only small items like cutlery and tools. Starting in January 1855 he began working on a way to produce steel in the massive quantities required for artillery and by October he filed his first patent related to the Bessemer process.

Bessemer first started working with an ordinary reverbatory furnace but during a test a couple of pig ingots got off to the side of ladle and were sitting above it in the hot air of the furnace. When Bessemer went to push them into the ladle he found that they were steel shells: the hot air alone had converted the outer parts of the ingots to steel. This crucial discovery led him to completely redesign his furnace so that it would force high-pressure air through the molten iron using special air pumps. Intuitively this would seem to be folly because it would cool the iron, but due to exothermic oxidation both the silicon and carbon react with the excess oxygen leaving the surrounding molten iron even hotter, facilitating the conversion to steel.

Bessemer licenced the patent for his process to five ironmasters, for a total of £27,000, but the licences failed to produce the quality of steel he had promised and he later bought them back for £32,500. He realized the problem was due to impurities in the iron and concluded that the solution lay in knowing when to turn off the flow of air in his process; so that the impurities had been burnt off, but just the right amount of carbon remained. However, despite spending tens of thousands of pounds on experiments, he could not find the answer.

The simple, but elegant, solution was first discovered by English metallurgist Robert Forester Mushet, who had carried out thousands of scientifically valid experiments in the Forest of Dean. His method was to first burn off, as far as possible, all the impurities and carbon, then reintroduce carbon and manganese by adding an exact amount of spiegeleisen. This had the effect of improving the quality of the finished product, increasing its malleability - its ability to withstand rolling and forging at high temperatures and making it more suitable for a vast array of uses.

The Dowlais Iron Company was the first licensee of the Bessemer process, constructing the world's most powerful rolling mill in 1857, and producing its first Bessemer steel in 1865.

The Bessemer process revolutionized steel manufacture by decreasing its cost, from £40 per long ton to £6-7 per long ton during its introduction, along with greatly increasing the scale and speed of production of this vital raw material. The process also decreased the labour requirements for steelmaking.

Prior to its introduction, steel was far too expensive to make bridges or the framework for buildings and thus wrought iron had been used throughout the Industrial Revolution. After the introduction of the Bessemer process, steel and wrought iron became similarly priced, and most manufacturers turned to steel. The availability of cheap steel allowed large bridges to be built and enabled the construction of railways, skyscrapers, and large ships.

Other important steel products were steel cable, steel rod and sheet steel which enabled large, high-pressure boilers and high-tensile strength steel for machinery which enabled much more powerful engines, gears and axles than were possible previously. With large amounts of steel it became possible to build much more powerful guns and carriages, tanks, armored fighting vehicles and naval ships. Industrial steel also made possible the building of giant turbines and generators thus making the harnessing of water and steam power possible. The introduction of the large scale steel production process perfected by the Englishman Henry Bessemer paved the way to mass industrialization as observed in the 19th-20th centuries.

Obsolescence

Commercial steel production using this method stopped in Workington in 1974. It was replaced by processes such as the basic oxygen (Linz-Donawitz) process, which offered better control of final chemistry.

The Bessemer process was so fast (10–20 minutes for a heat) that it allowed little time for chemical analysis or adjustment of the alloying elements in the steel. Bessemer converters did not remove phosphorus efficiently from the molten steel; as low-phosphorus ores became more expensive, conversion costs increased. The process permitted only limited amount of scrap steel to be charged, further increasing costs, especially when scrap was inexpensive. Use of electric arc furnace technology competed favorably with the Bessemer process resulting in its obsolescence.

A slag ladel outside Workington Steel Works in 1986. Up until the 1960's slag from blast furnaces was taken to the shore and dumped into the sea.

Betts Electrolytic Process

In Betts process lead bullion is purified by transferring most of the lead from a soluble anode to a cathode through a lead containing electrolyte while leaving behind impurities in an adherent anode slime. The electro- chemistry of this process involves the properties of anode slimes and of entrained electrolyte. Thus, the processes of anodic corrosion of lead and transport of lead ions through the entrained electrolyte are essential to the understanding of the Betts process.

Extractive Metallurgy of Lead

Lead is an ancient metal. It was used by the Romans for components of their water distribution systems because of its malleable and ductile properties. The ancients made lead of acceptable purity probably by smelting lead ores under conditions that prevented arsenic, antimony, and other hardening elements from reducing to the metallic phase. This could be accomplished in most cases by producing high lead slags such as those still produced in fire as saying. In those days it was necessary to avoid excessive copper in the ore, but arsenic, antimony, bismuth, nickel, iron, etc. were reliably held in the slag by maintaining the high oxidizing conditions of lead silicate based slags. Lead recoveries were tow not better than 85% from the highest grade handpicked galena ore. When high lead recoveries were found to be obtainable by coke-based blast furnace reduction, lead so produced was too hard, usuaily because of its copper, antimony, and arsenic content. The function of lead refining became both a softening process and a method of recovering silver and gold.

Nowadays the extraction process can be conveniently portrayed in the two step flow sheet shown in figure below. The first step involves bullion production from the sulphide concentrate and the second step the refining of bullion to the final product.

The conventional route for bullion production requires sintering of the concentrate to produce a lead oxide containing product, which is then reduced in a blast furnace with metallurgical coke to produce lead bullion. The KIVCET and QSL processes represent two relatively recent commercial developments that replace both the sintering-blast furnace combination with a single furnace that treats concentrates, and reduce both the costs of lead smelting and the environmental impact.

The KIVCET process replaces the sintering/blast furnace operations with a flash smelting step. In this process, lead sulphide is oxidized to lead bullion and sulphur dioxide in a stream of oxygen. In a second step, the bullion and slag flow under a weir to an electrically heated settling hearth where coke breeze or coal is added to reduce the residual lead oxide in the slag and produce a final bullion (for refining) as well as a low-lead stag.

The QSL process consists of a long, horizontal, tubular, brick-lined converter in which lead concentrates are pelletized and injected near one end into a bath containing lead bullion, lead oxide-containing slag, and lead sulphide matte. Oxygen is blown into the bath in the feed injection (and lead-bullion discharge) zone where it ultimately oxidizes sulphide sulphur to sulphur dioxide gas. Slag is tapped from the far end of the reactor after passing through a zone where reducing coal-air mixtures are injected through tuyeres to lower its lead oxide content.

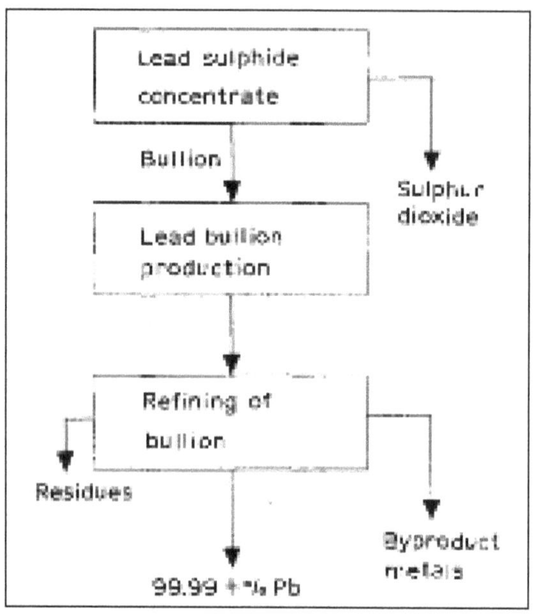

Figure: Flow sheet for lead extraction from sulphidic concentrates.

The refining of bullion is carried out by either a pyrometallurgical route or a combined pyrometallurgical/electrometallurgical route. Comparisons between these two routes show that the pyrometallurgical route is usually used when ores with low bismuth content are treated. The generalized pyrometal- lurgical flowsheet is shown in figure below. This process consists of a series of steps which capitalize on a complex series of phase relationships to extract all the impurities contained in the lead bullion down to very low levels.

The combined pyrometallurgical/electrometallurgical route is shown in figure below. Here, copper drossing is performed to remove the bulk of the copper as a combination of matte and arsenide-antimonide for further treatment, thus allowing for the removal of some arsenic and antimony. Arsenic and antimony are sometimes reduced further as sodium arsenate- antimonate dross by oxidizing in the presence of caustic soda, because their levels in bullion must be controlled to produce suitable anodes for successful electrorefin- ing practice.

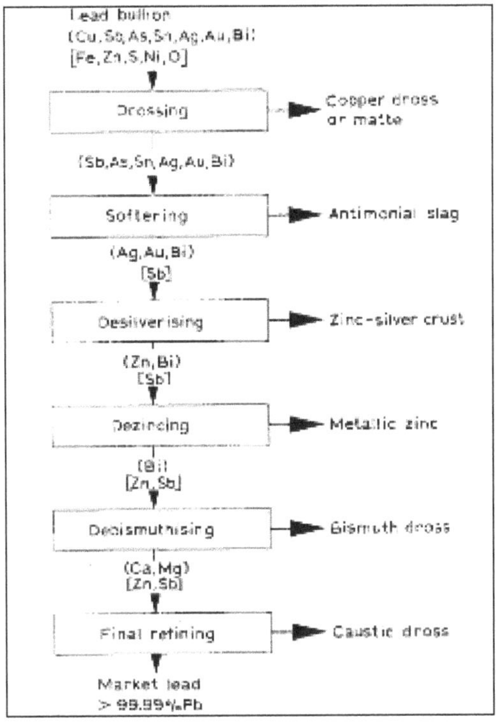

Figure: "A generalized flow sheet for the pyrometallurgical refining of lea d. Parentheses ( ) indicate a major impurity component. Brackets [ ] indicate a minor impurity component.

Electrorefining is carried out in either a fluosilicic, fluoboric or sutphamic acid electrolyte and produces a commercial lead cathode product and an anode with an adhering slime. The purity of the produced lead is usually higher than 99.99%. The slimes, representing only 2 to 4% of the anode weight, are treated by a variety of processes to recover silver, copper, antimony, gold, bismuth, and sometimes tin and indium.

Processes that entirely avoid smelting (and its attendant gas and dust treatment systems and associated environmental risks), utilizing hydrometallurgical/ electrometallurgical flowsheets, have also been proposed to replace the current technology. These include (a) the US Bureau of Mines ferric chloride leach process which recovers lead via the molten salt electrolysis of PbCI2, (b) the Minemet Recherche ferric chloride leach process which recovers lead from chloride leach solutions using aqueous electrolysis and (c) the US Bureau of Mines process for leaching lead concentrates in waste fluosilicic acid with an oxidant (hydrogen peroxide or lead peroxide) followed by aqueous electrolysis from the fluosilicic acid leach solution. All these processes have been piloted, but none has been commercialized.

Plant Practice in Lead Electrorefining

The BEP process is used in Canada, China, East Germany, Italy, Japan, Peru, Rumania, Russia, USA and West Germany. The average annual production of lead by BEP is approximately 1 000 000 tons. Since the production of refined lead in the non-socialist countries (for which good figures are available) is close to 4 700 000 tons per year, it can be assumed that up to about 20% of world lead production is refined by the BEP. Two variations of the BEP, the sulphamic acid and the ftuoboric acid processes are operated in Italy and in West Germany respectively.

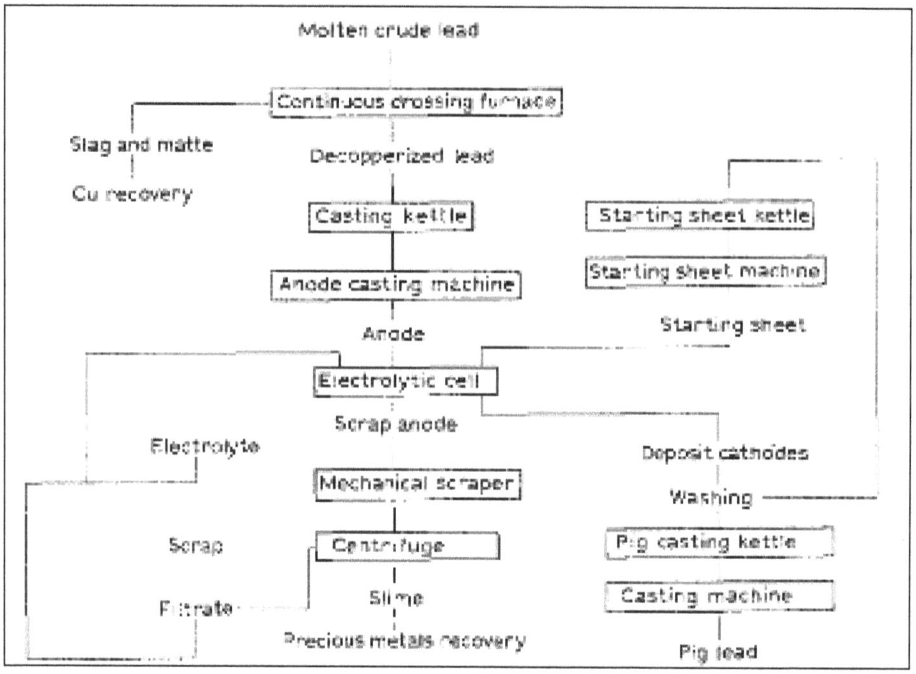

Figure: A generalized flowsheet for the pyrometallurgical/electrometalturgical refining of lead.

The Betts Electrorefining Process normally utilizes a fluosilicic acid (H2SiF6) electrolyte containing lead ttuositicate to electrorefine impure lead anodes into pure lead cathodes. The fluoborate and the sulphamate processes are identical to the fluosilicate process except for the substitution of the electrolyte. Fluoboric acid has not been extensively used due to its relatively high cost. HBF4 is a stable acid with a good electrolytic conductivity and a high solubility of the lead salt. It is also used in lead plating baths, lead fluoborate- fluoboric acid rechargeable batteries and in the recovery of lead from spent batteries.

The sulphamic acid process has two main limitations when compared to the fluosilicic acid process: firstly, sulphamic acid decomposes rapidly at current densities larger than 100 A m^{-2} resulting in high reagent replacement cost; secondly, the free acid is a crystalline solid of limited solubility, and limited ionization in aqueous solution, leading to a low con- ductivity of sulphamic acid solutions (at most, half of the conductivity of equivalent fluosilicic acid solutions). This results in larger power costs in refining. The significant advantage of the sulphamic acid process is that it is the most efficient process for the removal of tin from impure lead (Sn remains in the slimes). Plants that use the fluosilicic process remove Sn prior to electrolysis by using the Harris process. When anodes containing significant Sn concentrations are refined by the Betts process, tin dissolves with lead at the anode and co-deposits at the cathode. The lead-tin alloy may be sold, or some pos t treatment of the cathodes is necessary to remove the tin.

The wide use of H_2SiF_6 in lead electrorefining is due to its low cost. Fluosilicic acid is produced as a by-product of the treatment of phosphate rock in fertilizer manufacture. Fluorides and silica containing in phosphate rock form fluosilicate and are separated from the fertilizer product. Fluosilicic acid is also produced as a by-product during the dissol- ution of apatite with sutphuric acid.

During the electrorefining operation, fluosilicic acid is con- sumed by entrapment in the anode slimes and by volatilization from the surface of the electrolyte. The acid develops significant vapour pressures through volatile decomposition products according to the reaction,

$$H_2SiF_6 \rightarrow 2HF(g) + SiF_4(g)$$

SiF_4 and HF are both corrosive and toxic and are removed from the tankhouse atmosphere by adequate ventilation.

Higher current densities have been achieved through the use of galvanodynamic techniques such as current modulation and periodic current reversal (PCR). The current modulation technique consists of decreasing the current density (for example, from 220 to 160Am- 2) in small steps. Each constant current density step is determined on thebasis of an anode over potential value which increases as the slimes layer thickens and decreases when the current density is reduced. The upper limit for the anode over potential is usually determined by bismuth dissolution from the anode. Below figure shows the current density program that can be applied to the electrorefining circuit without reaching the critical over potential value for bismuth dissolution. The higher average current density possible with current modulation reflects in a shorter electrorefining cycle and higher refinery production.

Periodic current reversal (PCR) in lead electro- refining is widely employed in China. PCR involves frequent short reversals of the electrolysis current direction. This reduces the concentration polarization in the slimes layer and levels the cathodic deposit by selectivity dissolving projections. High current efficiencies, good cathode quality, low electrolyte losses, low energy consumption and a decrease in the number of short circuits has been reported through the use of PCR. A 16% increase in free acid was found in the slimes layer due to PCR.

Anodic Process

The Betts process, as practiced by all refineries, depends on the formation of an adherent, porous anode slimes layer during electrolysis. The slimes layer consists of undissolved impurities which are removed mechanically from anode scrap after the anodes are withdrawn.* If the slimes do not adhere to the anode 95'.

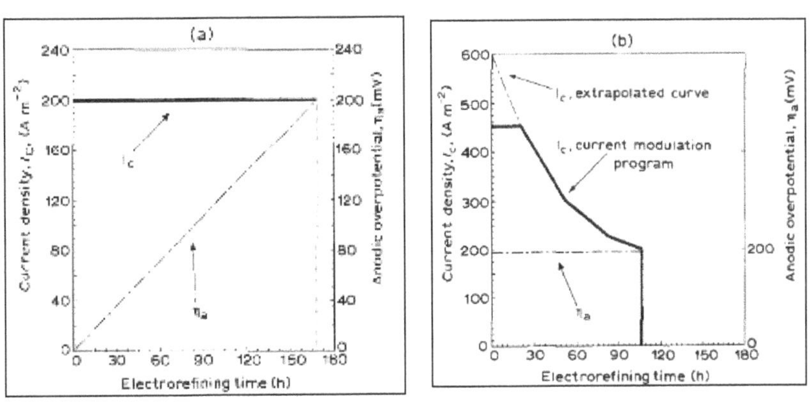

Figure: Changes in the anodic over potential value, η_a during lead electrorefining: (a) conventional galvanostatic process (b) galvanodynamic process in which the cell current is continuously decreased during the refining cycle to fix the value of η_a.

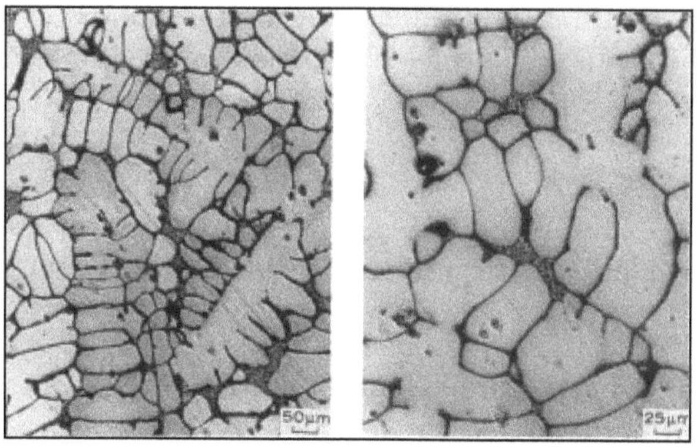

Lead anode microstructure. From an anode currently being used by Cominco Ltd. Chemical composition as described in Table1. Top view, air cooled side.

Refining Of Lead by the Betfs Process

During electrolysis they will settle through the solution and to some extent be mechanically entrained in the cathode deposit. There is no indication in the literature of a lead refining process in which (as in copper refining) slimes t:atl is encouraged. Further, the recovery of slimes from the bottom of the cell (as in copper refining) is perceived to be more costly. To form an adherent slimes layer, certain elements (mainly As, Sb, and Bi) must be controlled within a narrow composition range and/or ratio in the anode, and the anode casting process must be designed to control the rate of solidification to optimize the microstructure of the cast bullion.

Physical Metallurgy of the Lead Anodes

The lead anode microstructure that is desired for optimum slimes structure during electrorefining is known as the honeycomb structure, because it consists of uniform size grains of lead surrounded by impurities on the grain boundaries. As the lead grains dissolve they leave behind a skeleton of slimes which resembles a honeycomb. Any non-uniformity of the matrix will lead to non-adherence of the slimes layer, and any precipitates present in this matrix material will contribute to slimes detachment because of their extra weight.

There are four elements present in the lead anodes that seem to exert a strong influence on the anodic process: As, Sb, Bi and Ag. The interaction of these impurities with each other and with lead can be deduced from the available binary and ternary phase diagrams [89-93], which show that both intermetallic compounds and eutectic structures may be present. Three different slime forming systems have been identified and classified from studies on synthetic anodes:

(i) Impurity phases of the solid solution type (SST): the Pb-Bi system.

(ii) Impurity phases of the precipitation type (PT): the Pb-Sb system.

(iii) Impurity phases of the eutectic type (ET): the Pb-As and the Pb-Ag systems.

The strength of slimes adhesion to the anode was quantified by Tanaka through observations of slimes fall and slimes morphology. Tanaka summarized the results of these studies with the following relationships:

(1) The greatest slimes adherence is obtained when impurity phases of the SST type are present in the anode in concentrations greater than 0.23% by wt. A SST concentration lower than this critical value produced a slime that easily slides off the anode.

(2) The addition of a third element to the eutectic systems increases significantly the slimes adherence.

(3) Water quenching of the anodes increase slimes adhesion particularly in the PT and ET system cases. However, the very small slimes particles formed during quenching may also promote slimes detachment and mechanical entrainment in the cathodes.

The increase of slimes adhesion by water quenching relates directly to the lead anode solidification rate, which influences the growth and distribution of impurity-containing phases. Especially important are the eutectic forming systems where it is known that the solidification parameters (growth velocity, temperature gradient in the liquid and growth mechanism) and system parameters (volume fraction and impurities content) can lead to anomalous eutectic structures. Such structures have been reported to occur in the following systems:

Ag-Pb and Ag-Bi: broken lamellar structure type.

Pb-Bi: complex structure type.

Pb-Sb: complex regular structure and irregular structure type.

It must be emphasized that even though the physical metallurgy of the lead anodes is of great importance to the Betts process, there remains a lack of knowledge in this area.

Industrial Practice

The range of anode impurities used in lead refining varies from plant to plant. In every refinery the anode composition is kept within narrow limits to obtain an adherent slimes layer. The need for production of anodes with homogeneous properties has led to control methods for cooling rates and casting techniques. In addition to the use of a casting wheel, straight horizontal and vertical anode casting systems are employed,

The straight type horizontal lead anode casting system currently used in Japan. Although the vertical and horizontal casting processes were originally developed to save space (over that occupied by a casting wheel), they were carefully designed to achieve uniform cooling rates during the casting of the anodes. Such new techniques for anode casting as well as for optimizing heat treatment have been patented by Japanese companies , but no information was found that would indicate the nature of changes in microstructure resulting from these newer techniques.

Slimes Electrochemical Behavior

The slimes layer formed in the BEP undergoes structural and chemical changes during its growth. As the slimes layer thickens, the different phases and com- pounds present react with the entrained electrolyte creating secondary products. The Eh-pH diagram for the $PbSiF_6$-H_2SiF_6 system indicates that both lead fluoride and silica can precipitate at higher pH values. This has been related to the effect of electrolysis parameters on secondary processes (such as these precipitations) that take place within the slimes layer during electrolysis. The transport processes within the slimes layer in the context of these secondary reactions has also been the subject of several studies.

A simple physical model to study the role of the slimes layer during anodic dissolution rate in electro- refining systems was developed by Reznichenko. In this model, the slimes layer was represented by a silver gauze diaphragm (representing the noble impurity) electrically connected to the anode (a highly pure base metal) and placed at a (variable) distance from the anode corresponding to the slimes - electrolyte interface. The distance between the anode and the gauze was varied to simulate the effect of an increasing ohmic drop between the anode and the slimes layer. During electrolysis, concentration gradients normally present in a slimes layer were replaced by a simple concentration difference between the solutions on the two sides of the gauze, and this was related to a modified Nernst equation:

$$\text{In}\left[\text{Me}_2^{n_2^+}\right]=\left(E_1^{\text{o}}-E_2^{\text{o}}+\Delta\phi+\frac{RT}{n_1F}\text{In}\left[\text{Me}_2^{n_1^+}\right]\right)\frac{n_1F}{RT}$$

where: $\left[\text{Me}_2\right], n_2$ = concentration and valence of noble metal; $\left[\text{Me}_1\right], n_1$ = concentration and valence of base metal; $\Delta\phi$ = ohmic drop between the base metal and the noble metal gauze; E_1^{o} = rest potential of the metal to be refined; E_2^{o} = est potential of the noble metal; F = Faraday's constant; R = gas con- stant; and T = temperature.

Equation $\text{In}\left[\text{Me}_2^{n_2^+}\right]=\left(E_1^{\text{o}}-E_2^{\text{o}}+\Delta\phi+\frac{RT}{n_1F}\text{In}\left[\text{Me}_2^{n_1^+}\right]\right)\frac{n_1F}{RT}$ shows that the larger the ohmic drop the larger the equilibrium concentration of ions of the noble metal and the larger its dissolution rate. Although this model oversimplifies the different phenomena taking place within a slimes layer, it pro- rides an insight as to the effect of ohmic drop buildup in the slimes layer and the dissolution rate of noble impurities normally left in the slimes. Reznichenko. applied their model to the study of the Cu-Ag system in which they found a logarithmic relationship between the dissolution rate of noble impurities and the ohmic drop potential.

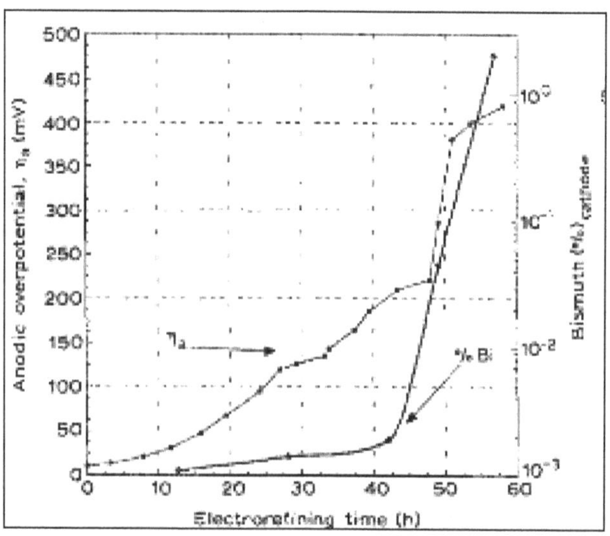

Figure: Bismuth content of the cathodic deposit as a function of the anodic polarization and the electrolysis time. Ic = t70-350 A m -2. Anode composition 8.4-11.2% Bi, remainder Pb.

Early research on the behavior of BEP slimes has focused on relating the anodic overpotential to the cathode purity. Among the impurities most closely followed in the cathodic deposit is bismuth. Even though the BEP has a large selectivity for the removal of bismuth, it has been found that towards the end of the electrorefining cycle such selectivity can be lost. It seems that the buildup of concentration gradients and the precipitation of secondary products within the slimes layer can initiate steep increases or discontinuities in the anodic over- potential at a certain slimes thickness. This would produce a large increase in the rate of dissolution of such impurities, and subsequently in the deposition of impurities in the cathodic deposit. Figure above shows how bismuth contamination in the cathode increases with anodic overpotentials above a critical value (about 200 mV). This has been recognized as a general behavior for bismuth in lead refining by the BEP. On the other hand, the use of periodic current reversal (PCR) has been found to increase the minimum anodic overpotential at which noble impurities start to dissolve.

The slimes layer weight is only 1 to 4% of the original anode weight; yet, they occupy the whole of the original anode volume. This indicates that the porosity of the slimes layer exceeds 92% and may be as high as 98% (taking into account a density for slimes phases of about half that of the bullion). Concentration gradients in the electrolyte confined within this highly porous layer and the slimes electrochemistry are closely linked.

Wenzel measured the change in composition of the electrolyte contained in the slimes layer as a function of time and of slimes thickness. Figure shows the sampling method used by Wenzel, utilizing sampling wells at different distances from the anode surface. Small amounts of electrolyte were withdrawn through these wells at carefully selected times (to avoid perturbing appreciably the system).

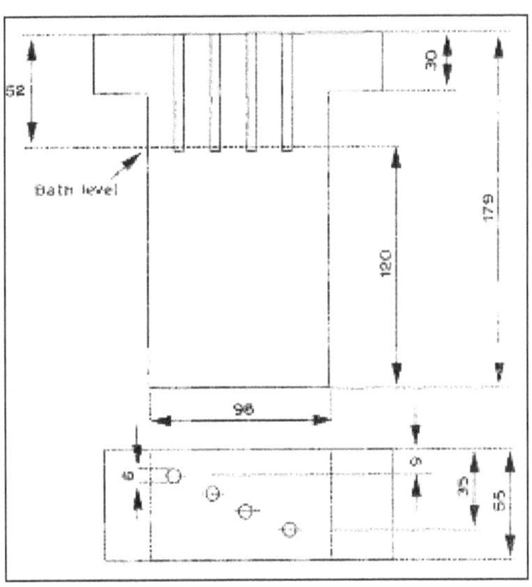

Figure: Dimensions of the anodes used in Wenzel's experiments showing the position and size of the electrolyte sampling wells. All measurements in ram.

According to their results, using a range of anode compositions (Bi from 0 to 1.74% and Sb from 0.45 to 3.01%), the more the amount of secondary solidified material (SSM) present in the anode, the steeper the pH and the Pb^{+2} concentration gradient throughout the slimes layer. Also they found that the thicker the slimes layer the steeper the concentration gradients. Figure shows these changes in concentration for two different anode compositions. Wenzel et al. proposed that due to the pH increase several reactions may occur: SiF_6^{-2} decomposition to SiF_4 and F (pH > 3); precipitation of Sb203 (pH > 4.9) and precipitation of Pb_2O (pH > 7).

Table below show the results of chemical and diffraction analysis, which seems to indicate the presence of these secondary compounds within the slimes layer*. From an analysis of the concentration of noble impurities in the entrapped electrolyte, Wenzel were able to conclude that when the SSM exceeded 5%, as determined from the Pb-Bi-Sb ternary diagram, permissible impurities in the anode were too high to obtain an acceptable impurity level in the refined lead.

Wenzel's X-ray diffraction analysis on the slimes pro- ducts agree with the findings of Isawa et al., who studied the slimes composition of synthetic and industrial lead anodes and found the presence of the following compounds:

Metallic Bi: in the Pb-Bi, Pb-Bi-As, Pb-Bi-Sb, and Pb-Bi-As-Ag systems;

Metallic As: in the Pb-As, Pb-Bi-As, and Pb-Bi- As-Ag systems;

Metallic Ag: in the Pb-Ag and Pb-Bi-As-Ag systems.

Chemical composition and X-ray diffraction analysis of the lead anode slimes.

Sample no.	Anode composition (% wt) (rest Pb)	Slimes chemical analysis (% wt)					X-ray diffraction analysis
		Pb	Sb	Bi	F	Si	
1	0.46% Sb, 0.24% Bi	56.98	7.62	3.95	11.97	6.66	PbO, Sb_2O_3, PbF_2, Bi_2O_3
2	0.46% Sb, 0.24% Bi	58.02	9.40	4.28	14.09	3.33	PbO, Sb_2O_3, PbF_2, Pb_4SiO_4, Bi_2O_3
3	0.92% Sb, 0.24% Bi	56.57	6.81	5.55	11.27	6.65	PbO, Sb_2O_3, PbF_2, Bi_2O_3
4	0.92% Sb, 0.24% Bi	53.77	9.37	9.20	12.22	4.19	PbO, Sb_2O_3, PbF_2, Bi_2O_3

Also the presence of the Bi-Sb solid solution, of the ε and ε' phases of the Ag-Sb system, and of water soluble As and Sb (identified as As203 and of Sb203) were found in some of the above mentioned systems and in the slimes obtained from industrial anodes.

One compound that has been difficult to characterize is Ag3Sb , whose presence has been detected in both synthetic and industrial lead anodes. The morphology of this compound has been found to be a function of the anode cooling rate and of the electrolysis conditions.

The electrolyte used in lead electrorefining contains impurities whose concentration ranges from a few ppm and several gl^{-1}, and have their origin in the anode from which they are dissolved (i.e. C_u +2, S_n +2) or from the acid manufacturing process (i.e. phosphorous species). Their steady state concentrations in the electrolyte depend on the electrolysis parameters and slimes thickness. They can affect both the anodic and cathodic processes through changes in fundamental electrochemical parameters such as the exchange current density (i_0), the transfer coefficient (α), and the electrical double layer capacity (e.d.1.). Measurement of these parameters usually involve transient electrochemical techniques, such as polarization scans, cur- rent interruptions, and AC impedance studies. Miyashita have extensively studied the influence of minor impurities and addition agents in the electrolyte used in lead refining. These studies focused on the determination of fundamental electro- chemical parameters (i_0, e.d.1., and α) as a function of Sn^{-2},Fe^{+3},Zn^{+2} and glue, using a single current step transient method. He found that under an oxygen atmosphere the presence of the above ionic species increased the anodic overpotential value. Thus, they may inhibit the dissolution of some noble species including lead.

When noble impurities present in the lead anodes transfer to the electrolyte they can be redeposited by cementing on less noble elements. While such redeposition is obscured during anodic dissolution, it can be inferred from corrosion measurements in the absence of a current. For example, the weight loss rate for a lead sample in the presence of dissolved bismuth has been observed in the HBF_4-$Pb(BF_4)$ 2 system. Figure below shows how bismuth concentrations as low as 2 mM enhance the corrosion rate of lead.

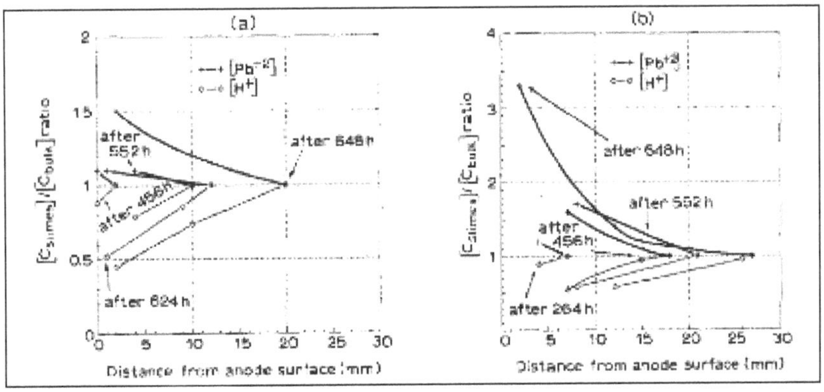

Figure: Changes with electrolysis time of the relative concentrations of Pb^{+2} and H + within the slimes layer with respect to their bulk values.

Bulk electrolyte composition: [$PbSiF_6$] = 0.4 M, [H2SiF6] = 1 M, T = 35° C, I = 200 a m -2, stationary electrodes. Vertical axis: Pb $^{+2}$ and H^+ concentration ratio between the electrolyte sampled within the slimes layer and the bulk electrolyte. (a) Anode with 0.46% Sb and 0.24% Bi (~ 1.73% SSM), (b) anode with 0.92% Sb and 0.24% Bi (3.80% SSM). Note: no extrapolation of the inner electrolyte concentrations at zero slimes thickness done here.

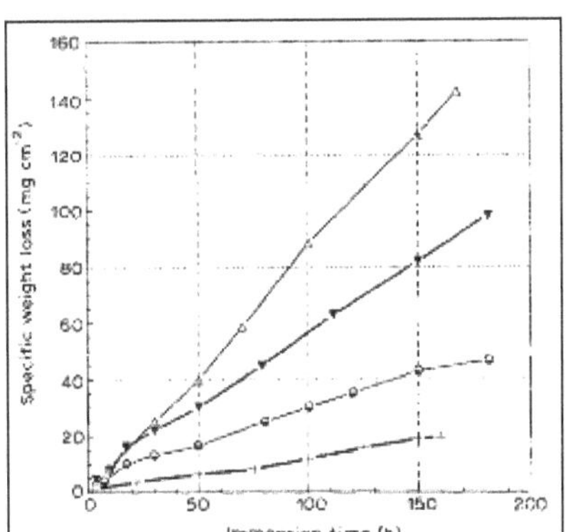

Figure: Lead specific weight loss (corrosion) as a function of the immersion time in a 2 M Pb $(BF_4)_2$ - 1 M HBF 4 solution in the presence of(+) 0, (Θ) 2, (∇) 5, and (Δ) 10Mm1⁻¹ofBi $^{+3}$.

During the dissolution of the lead anodes in the BEP, bismuth and other noble impurities accumulate in the proximity of the slimes-bulk electrolyte interface as well as in the bulk electrolyte. Since these noble impurities will deposit on the cathode at their limiting mass transfer rate, the permitted electrolyte concentration is related to cathode purity specifications (25 mg 1-i of bismuth will typically lead to 10 ppm bismuth in the cathode). They can be removed by one of two methods:

(a) Continuous cementation through a column filled with lead particles.

(b) Electrodeposition in a separate electrolysis circuit. The purified electrolyte is then sent to the main electrolyte stream. This method is being used in Japan.

The majority of plants that electrorefine lead do not purify their electrolyte, but rather, operate under conditions where the dissolution of noble impurities from the anode is limited to tolerable levels.

Cathodic Process

Additives Control and Electrochemistry

In the absence of additives, lead deposits with a very small overpotential, and tends to form rough, porous deposits or dendrites that result in short circuits. To obtain deposits that are flat, smooth, and free from projections, special reagents are added to the electrolyte. These 'addition agents' increase the cathodic overpotential (actually called 'inhibition'), and change the kinetic parameters ($i_{o,\alpha}$ e.d.l.) under comparable electrolysis conditions.

Polarization measurements have been used for con- trolling and monitoring the concentration of additives in lead electrorefining circuits, as welt as for screening of additives in the electrolytic refining of copper and for testing of impurity levels in the electrowinning of zinc. A polarization technique for determining lignin sulphonate in lead plating baths has also been described. Long term studies and years of industrial practice, especially by Cominco researchers have lead to a sufficient under- standing of the levelling mechanisms of lignin su!- phonate and aloes, to permit the use of polarization measurements for controlling the concentration of these species in the industrial electrolyte for lead refining. This additives control is achieved by keeping the cathode polarization voltage (c.p.v.) within preset limits, and adjustments are made by either new additions or by regeneration (in the case of aloes) with a thiosulphate salt. The level of additives present at a given time is directly related to the c.p.v.

Starting Sheet Technology

The cathode starting sheets used in the BEP are usually cast on a rotary drum casting machine. The stiffness of these sheets is increased by adding approximately 15 ppm antimony to the melt and by impressing Continuous purification of electrolyte via purification coaumn wrinkles on the cathodes right after they are produced. Procedures to avoid dross incorporation in the lead cathodes have also been developed. In this topic of the electrorefining process high levels of automation have been achieved.

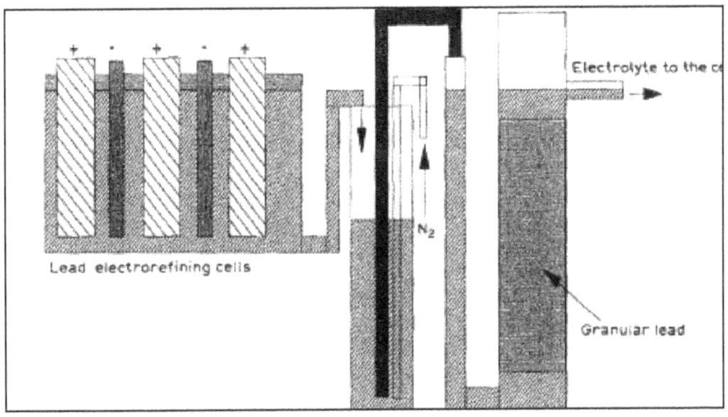

Figure: Continuous purification of electrolyte via purification columm.

Cell Electrolysis Parameter Optimization

The overall electrorefining process has been studied using statistical correlations of some of the variables measurable in an operating plant. To optimize the electrolysis parameters when the BEP is run at high current densities $(\gg 200Am^{-2})$ actorial design of experiments at three levels for four variables has been used by Lange. They varied PbSiF6 and HaSiF6 concentrations, temperature, and current density, at constant values of addition agent (glue), anode composition, cell geometry, and electrolyte recirculation. The influence of these parameters on the average anodic and cathodic polarization, on the cell voltage, and on the specific energy consumption was obtained by employing regression analysis correlations. In addition, changes in these parameters were correlated with the cathode quality and appearance as well as with the consumption of glue. The optimum electrolysis parameters found for a current density increase from 200 to 300 Am^{-2}) were a Pb +2 concentration of 60 g 1^{-1} (down from 80 g 1^{-1}), a glue addition maxi- mum of 1500gton of lead (up from 1250g1^{-1}), a free H_2SiF_6 content of 120gl^{-1} (no change) and an electrolyte temperature of 35 to 40° C. Under these electrolysis conditions, Lange et al. found that antimony and bismuth slimes layers could be subjected to anodic polarizations as high as 280mV without serious impairment of the cathode quality.

Optimization of the BEP through half cycle anode slimes scrubbing and cathodes exchange was also investigated by Lange. Not only did the specific energy consumption decrease by the implementation of this procedure, but highly pure cathodes were assured by avoiding anodic overpotentials in excess of the preset limits for impurities dissolution.

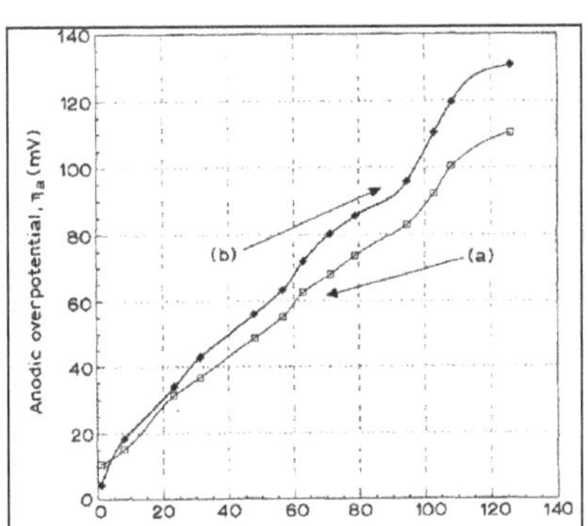

Figure: Influence of the cycle length and current reversat ratio on the anodic overpotential value during PCR: $(a)T_{forward} / T_{backward} = 11.5, T_{cycle} = 10s$. $(b)T_{forward} / T_{backward} = 19, T_{cycle} = 20s$.

On the other hand, the exchange of cathodes and the half cycle scrubbing of the anodes incorporates labor increases making implementation difficult. The use of this technique is necessary when anodes containing high impurity levels (> 5%) are to be treated. For example, in Russia, lead anodes containing as much as 15% bismuth are scrubbed every 48 h during the 6 day long anode cycle. In addition to this, cathodes are exchanged daily to avoid short circuits and to obtain a deposit of acceptable purity.

Lange et al. also studied the use of PCR in the BEP. Anodic overpotential dependance on the PCR parameters. Thiet found that the use of PCR decreased the anodic overpotential values. The reduction in the anodic overpotential values did not reflect in lower PCR energy consumption probably because it was offset by unproductive energy consumption during the reverse current phase of the cycle. Thiet also showed that the quality of the PCR cathode deposits was very high.

Bipolar Refining of Lead

Among other alternatives to optimize the lead electro- refining process, the use of a bipolar configuration looks most promising. In this configuration only the terminal electrodes are connected to the source of current. Between these electrodes a large number of bipolar electrodes can be incorporated. One side of these electrodes will corrode anodically, while pure lead will deposit on the other side. The impurities will be left behind as a slimes layer on the anodic side. The advantages of operating the BEP in a bipolar mode include PCR, c.p.v., optimum lead and free acid con- tents, and the use of jumbo electrodes (~ 4 m 2, possible because there are no bus-bar connections)*. The process has been operated in a pilot plant where it proved to be superior to the parallel process. The rationale for the current use of the parallel system is that a high economic investment is required for the substitution and implementation of the bipolar process. If new plants to electrorefine lead are to be constructed in the future, very likely they will be assembled on the bipolar configuration.

Patio Process

Patio process is also called Mexican process. It is the method of isolating silver from its ore that was used from the 16th to early in the 20th century; the process was apparently commonly used by Indians in America before the arrival of the Europeans.

The silver ore was crushed and ground by mule power in arrastras, shallow circular pits paved with stone. Large blocks of stone attached by beams to a central rotating post were dragged around the arrastra, reducing the ore to a fine mud. This was then spread over a courtyard or patio, sprinkled with mercury, salt, and copper sulfate, and mixed by repeatedly driving mules over it. Chemical reactions freed the silver from its compounds and caused it to dissolve in the mercury. When the amalgamation was complete, the material was agitated with water in large tubs and the mud run off. The amalgam remaining at the bottom was collected and heated to drive off the mercury. The process, especially suitable for the silver ores of the dry, barren areas of Mexico, was responsible for a large proportion of the world's silver production for 350 years; it was finally displaced by the cyanide process early in the 20th century.

Development of the Patio Process

Bartolomé de Medina was a successful Spanish merchant who became fascinated with the problem of decreasing silver yields from ores mined in Spanish America. By the mid-sixteenth century, it was well known in Spain that American silver production was in decline due to the depletion of high-grade ores and increasing production costs. The New Laws, prohibiting the enslavement of Indians, had resulted in higher labor costs as miners turned to wage labor and expensive African slaves. These higher production costs made mining and smelting anything but the highest grade silver ores prohibitively expensive, just as the availability of high grade ores was in decline. Bartolomé de Medina initially focused his attentions on learning about new smelting methods from smelters in Spain. He was approached during his research by an unknown German man, known only as "Maestro Lorenzo," who told him that silver could be extracted from ground ores using mercury and a salt-water brine. With this knowledge, Medina left Spain for New Spain (Mexico) in 1554 and established a model patio refinery in order to prove the effectiveness of the new technology. Medina is generally credited with adding "magistral" (a copper sulfate) to the mercury and salt-water solution in order to catalyze the amalgamation reaction. However, some historians assert that there were already sufficient copper sulfates in the local ores and that no additional magistral was needed. Regardless of whether or not Medina's contribution was entirely original, he promoted his process to local miners and was able to obtain a patent from the Viceroy of New Spain. As a result, he is generally credited with the invention of silver amalgamation in the form of the patio process.

Basic Elements of the Patio Process

Patio Process at the Gould & Curry Mill, Comstock Lode, USA, 1866.

Silver ores were crushed (typically either in "arrastras" or stamp mills) to a fine slime which was mixed with salt, water, magistral (essentially an impure form of copper sulfate), and mercury, and spread in a 1-to-2-foot-thick (0.30 to 0.61 m) layer in a patio, (a shallow-walled, open enclosure). Horses were driven around on the patio to further mix the ingredients. After weeks of mixing and soaking in the sun, a complex reaction converted the silver to native metal, which formed an amalgam with the mercury and was recovered. The amount of salt and copper sulfate varied from one-quarter to ten pounds of one or the other, or both, per ton of ore treated. The decision of how much of each ingredient to add, how much mixing was needed, and when to halt the process depended on the skill of an *azoguero* (English: quicksilver man). The loss of mercury in amalgamation processes is generally one to two times the weight of silver recovered.

The patio process was the first form of amalgamation. However, it is unclear whether this process or a similar process—in which amalgamation occurred in heated vats rather than open patios—was the predominant form of amalgamation in New Spain, as the earliest known illustration of the patio process dates from 1761. There is substantial evidence that both processes were used from an early date in New Spain, while open patios were never adopted in Peru. Both processes required that ore be crushed and refiners quickly established mills to process ore once amalgamation was introduced. By the seventeenth century, water-powered mills became dominant in both New Spain and Peru.

Due to amalgamation's reliance upon mercury, an expansion of mercury production was central to the expansion of silver production. From shortly after the invention of mercury amalgamation to the end of the colonial period, the Spanish crown maintained a monopoly on mercury production and distribution, ensuring a steady supply of royal income. Fluctuations in mercury prices generally resulted in corresponding increases and decreases in silver production.

Broader Historical Significance

The introduction of amalgamation to silver refining in the Americas not only ended the mid-sixteenth century crisis in silver production, it also inaugurated a rapid expansion of silver production in New Spain and Peru as miners could now profitably mine lower-grade ores. As a result of this expansion, the Americas became the primary producer of the world's silver, with Spanish America producing three-fifths of the world's silver supply prior to 1900.

While a number of factors resulted in the minimal use of forced Indian labor in Mexican silver production, the introduction of silver amalgamation allowed for an expansion of silver production in Peru that had profound consequences for Peru's native population. Francisco Toledo, the Viceroy of Peru in the 1570s, saw amalgamation as the key to expanding Peruvian silver production. He encouraged miners to adopt amalgamation and construct mercury mines. More significantly, in order to provide sufficient labor to accommodate the expansion of silver mining to lower-grade ores, Toledo organized an Indian draft labor system, the mita. Under this system, thousands of natives were forced to work in silver and mercury mines for less than subsistence-level wages. Twelve thousand draft laborers regularly worked at the largest mine in the Americas, located at Potosí in modern Bolivia. Native attempts to avoid the mita led to the abandonment of many Indian villages throughout Peru. Spanish monopolization of refining through amalgamation cut natives out of what had earlier been a native-dominated enterprise. Refining represented the most profitable segment of silver production. In conjunction with the mita, the exclusion of natives from owning refineries contributed to the transformation of Peruvian natives into a poorly paid labor force.

The rapid expansion of silver production and coinage—made possible due to the invention of amalgamation—has often been identified as the primary driver of the price revolution, a period of high inflation lasting from the sixteenth to early seventeenth-century in Europe. Proponents of this theory argue that Spain's reliance on silver coins from the Americas to finance its large balance of payments deficits resulted in a general expansion of the European money supply and corresponding inflation. Critics of the theory, however, argue that inflation was really a result of European government policies and population growth.

While the role of the expansion in silver production in the price revolution may be disputed, this expansion is often acknowledged as a key ingredient in the formation of early-modern world trade. Spanish American production fed Chinese demand for silver, facilitating the development of extensive trade networks linking Europe, Africa, Asia, and the Americas as Europeans sought to gain access to Chinese wares.

Pidgeon Process

The process involves reducing molten magnesium oxide slag by ferrosilicon under low gas pressure at a temperature of about 1400°C.

The metallic magnesium, formed in the process, evaporates and then condensates away from the hot region.

The condensed magnesium, having purity of 99.95% is then remelt and cast.

The Three Steps of the Pidgeon Process:

1. Calcination of Dolomite, at 1300° C, Highly Endothermic Reaction, CO2 Released both from the reaction, and from Fuel Burned for Process Heat.

   $CaMg(CO_3)_2 \rightarrow CaO + MgO + 2CO_2$

2. Ferrosilicon Alloy Production, the Ferrosilicon Alloy Production by Electric Arc through Mixture of Hematite, Quartz Sand, and Coal. Extremely Endothermic Reaction; Emits Toxic CO.

   $Fe_2O_3 + 4SiO_2 + 11C \rightarrow 2(Fe)Si_2 + 11CO$

3. Silicothermic Reduction of MgO by Ferrosilicon, at ~14000 C under Vacuum for 8- 10 hrs. Highly Endothermic Reaction, the ratio of feed are 6 calcine dolomite: 1 ferrosilicon as briquettes.

   $2MgO + 2CaO + (Fe)Si \rightarrow 2Mg(g) + Ca_2SiO_4(s) + Fe$

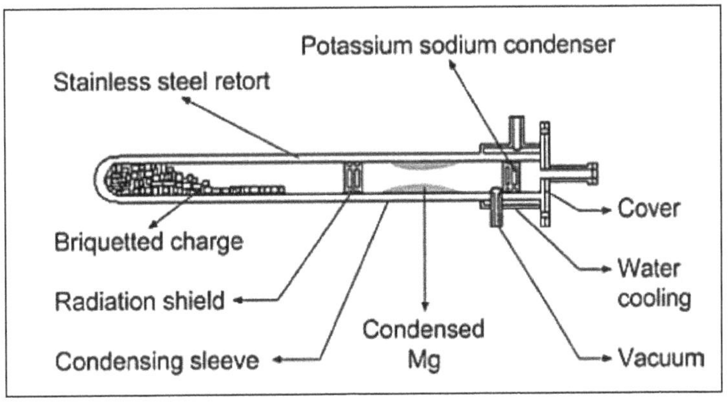

Hoopes Process

Hoopes process is mainly used on primary aluminium to produce extremely pure aluminium. This process is structured so that it consists of three liquid layers, which are separated due to different densities of the layers. This is illustrated in figure below. The bottom layer consists of an aluminium copper alloy mixed with the impure aluminium, and serves as the anode. The scrap is mixed with the copper to make it denser. The top layer consist of pure aluminium and serves as the cathode. These two layers are separated by a layer of molten electrolyte. The mid layer can consist of different substances, but normally this is a layer of molten inorganic salts, usually a chloride or fluoride. The idea of this process is to supply sufficient energy for the aluminium to react with the electrolyte and thus be transferred from the bottom layer up to the middle layer. When this happens, all elements more reactive than aluminium will transfer as well. Then the electrolyte is composed of components which form more stable compounds with the elements more reactive than aluminium, so only pure aluminium is able to travel up to the top layer. This way pure Al^{3+} is transported all the way to the top layer and can be removed as very pure aluminium.

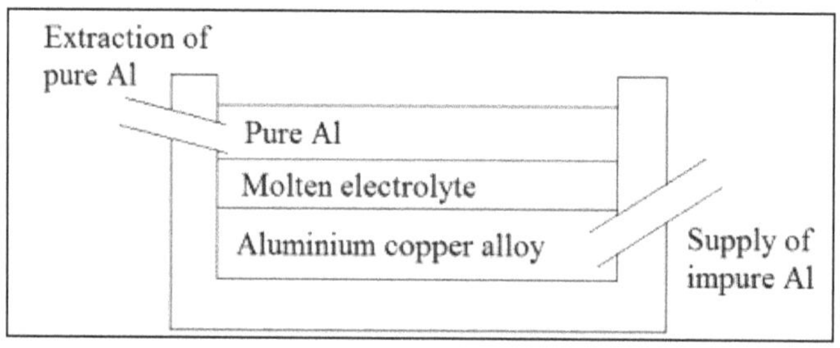

Figure: Principles of three-layered electrolysis.

The temperature requirement for this process is 700-900°C, and the energy use is between 17- 18 kWh/kg. The process can produce aluminium with a purity of >99.07 %. It is assumed that this energy requirement include both energy to reach desired temperature, to maintain this temperature throughout the process and the actual energy consumed by the process.

Schwartz and Wendt states that this process is able to achieve a purity level comparable to primary aluminium even for scrap with a purity level as low as 70 %.

The electrolyte and copper alloy used for this process can be reused and is thus not consumed during this process. They do however require some processing to get rid of the trapped impurities before they can be reused.

Adjusted Three-layered Electrolysis

As mentioned the three-layered electrolysis is mainly applied to primary aluminium to produce extremely pure aluminium. Kjos conducted an experiment to see if the three-layered electrolysis could be adjusted to fit aluminium scrap and possible reduce the energy requirements if the purity demands of the refined aluminium were lowered. They tested redesign of the cathode to reduce losses through voltage drop, and a new method redesigning the setup of the electrolysis. This is from a confidential report, and therefore only the energy use achieved can be included here. An energy use of respectively 8.5 and 5 kWh/kg were achieved.

Dow Process

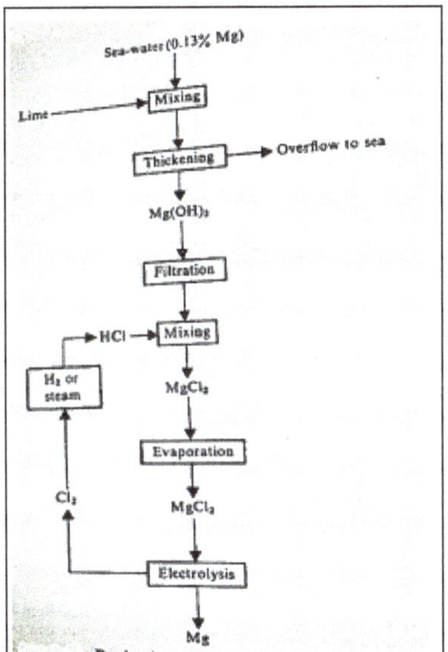

Production of Magnesium from Sea-Water.

The Dow Process is generally applied for extraction of Magnesium from Sea water. In this process first we add lime for thickening then mix with 10% HCl.

The electrolytic cell consists of a brick-lined vessel, divided into anode and cathode compartments by a semi-wall. Air or watercooled Graphite plate anode and steel cathode are submerged in electrolyte composed of alkaline chlorides with addition of magnesium chloride. The operating temperature is 680°C to 750°C. Magnesium chloride decomposes in the electrolytic cell according to the reaction.

$$MgCl_2 \rightarrow Mg + Cl_2\,(g)$$

Metallic magnesium is formed at the cathode. It floats up (it is lighter than electrolyte) collecting in the cathode compartment. Chlorine, which is by-product of the process, is collected in the anode compartment.

In this method CaCl2 and NaCl are added to anhydrous MgCl2 to decrease its melting point.

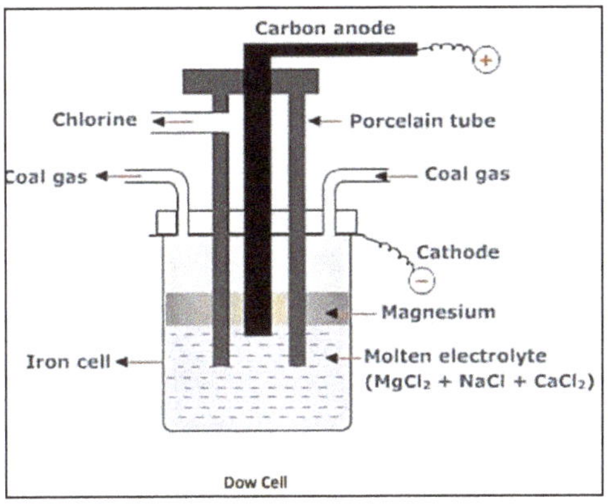

Dow Cell

References

- Dennis Flynn and Arturo Giráldez, "Born with a 'Silver Spoon': The Origin of World Trade in 1571," Journal of World History, Vol. 6, No. 2 (Fall, 1995), 202-203
- The-Bessemer-Process: wirralmodelengineeringsociety.co.uk, Retrieved 16 April 2020
- Patio-process, technology: britannica.com, Retrieved 30 March 2020
- Douglas Fisher, "The Price Revolution: A Monetary Interpretation," The Journal of Economic History, Vol. 49, No. 4 (Dec. 1989), 883-884
- The-refining-of-lead-by-the-betts-process: slideheaven.com, Retrieved 12 May 2020

5

Diverse Techniques in Metallurgy

Metallurgical techniques have been developed and improved over time to increase yield and efficiency while reducing costs and material losses. A diverse assortment of processes are used in metallurgy today, a few of them being mining, casting, welding, annealing, calcination, roasting, molding, electroplating, etc. This chapter has been carefully written to provide a better perspective of this topic.

Mining

Mining is the extraction of valuable minerals or other geological materials from the earth from an orebody, lode, vein, seam, reef or placer deposits which forms the mineralized package of economic interest to the miner.

Surface coal mining

Ores recovered by mining include metals, coal, oil shale, gemstones, limestone, dimension stone, rock salt, potash, gravel, and clay. Mining is required to obtain any material that cannot be grown through agricultural processes, or created artificially in a laboratory or factory. Mining in a wider sense includes extraction of any non-renewable resource such as petroleum, natural gas, or even water.

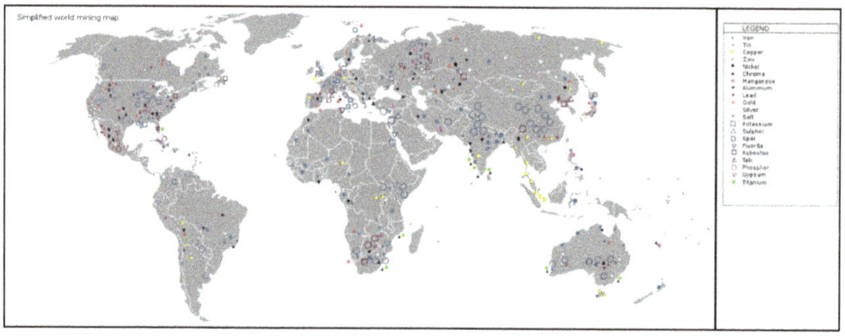

Simplified world active active mining map.

Sulfur miner with 90 kg of sulfur carried from the floor of the Ijen Volcano (2015).

Mining of stones and metal has been a human activity since pre-historic times. Modern mining processes involve prospecting for ore bodies, analysis of the profit potential of a proposed mine, extraction of the desired materials, and final reclamation of the land after the mine is closed.

Mining operations usually create a negative environmental impact, both during the mining activity and after the mine has closed. Hence, most of the world's nations have passed regulations to decrease the impact. Worker safety has long been a concern as well, and modern practices have significantly improved safety in mines.

Levels of metals recycling are generally low. Unless future end-of-life recycling rates are stepped up, some rare metals may become unavailable for use in a variety of consumer products. Due to the low recycling rates, some landfills now contain higher concentrations of metal than mines themselves.

History

Prehistoric Mining

Since the beginning of civilization, people have used stone, ceramics and, later, metals found close to the Earth's surface. These were used to make early tools and weapons; for example, high quality flint found in northern France and southern England was used to create flint tools. Flint mines have been found in chalk areas where seams of the stone were followed underground by shafts and galleries. The mines at Grimes Graves are especially famous, and like most other flint mines, are

Neolithic in origin (ca 4000 BC-ca 3000 BC). Other hard rocks mined or collected for axes included the greenstone of the Langdale axe industry based in the English Lake District.

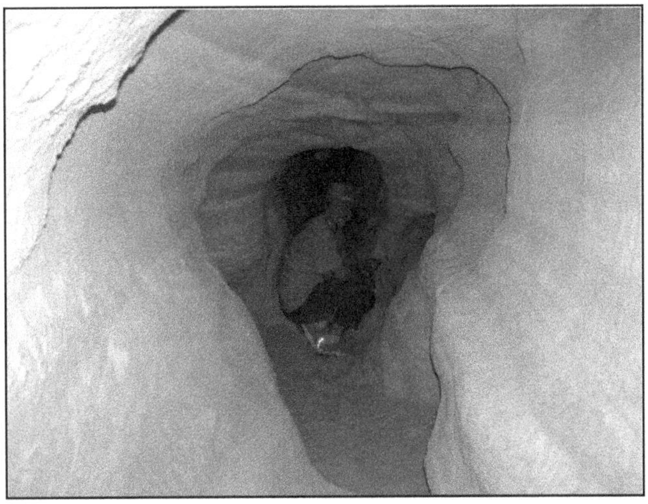

Chalcolithic copper mine in Timna Valley, Negev Desert.

The oldest known mine on archaeological record is the "Lion Cave" in Swaziland, which radiocarbon dating shows to be about 43,000 years old. At this site Paleolithic humans mined hematite to make the red pigment ochre. Mines of a similar age in Hungary are believed to be sites where Neanderthals may have mined flint for weapons and tools.

Ancient Egypt

Ancient Egyptians mined malachite at Maadi. At first, Egyptians used the bright green malachite stones for ornamentations and pottery. Later, between 2613 and 2494 BC, large building projects required expeditions abroad to the area of Wadi Maghareh in order to secure minerals and other resources not available in Egypt itself. Quarries for turquoise and copper were also found at Wadi Hamamat, Tura, Aswan and various other Nubian sites on the Sinai Peninsula and at Timna.

Mining in Egypt occurred in the earliest dynasties. The gold mines of Nubia were among the largest and most extensive of any in Ancient Egypt. These mines are described by the Greek author Diodorus Siculus, who mentions fire-setting as one method used to break down the hard rock holding the gold. One of the complexes is shown in one of the earliest known maps. The miners crushed the ore and ground it to a fine powder before washing the powder for the gold dust.

Ancient Greek and Roman Mining

Mining in Europe has a very long history. Examples include the silver mines of Laurium, which helped support the Greek city state of Athens. Despite the mine having over 20,000 slaves working in them, the technology was essentially identical to their Bronze Age predecessors. Other mines, such as on the island of Thassos, had marble quarried by the Parians after having arrived in the 7th Century BC. The marble was shipped away and was later found by archaeologists to have been used in buildings including the tomb of Amphipolis. Philip II of Macedon, the father of Alexander the Great, captured the gold mines of Mount Pangeo in 357 BC to fund his military campaigns. He also captured gold mines in Thrace for minting coinage, eventually producing 26 tons per year.

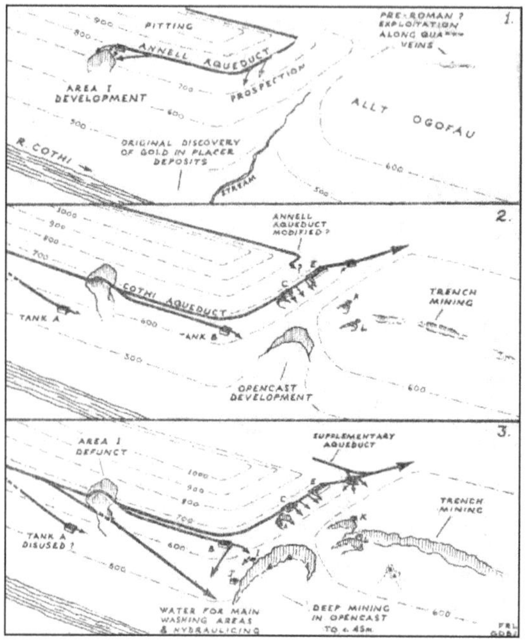

Ancient Roman development of the Dolaucothi Gold Mines, Wales.

However, it was the Romans who developed large scale mining methods, especially the use of large volumes of water brought to the minehead by numerous aqueducts. The water was used for a variety of purposes, including removing overburden and rock debris, called hydraulic mining, as well as washing comminuted, or crushed, ores and driving simple machinery.

The Romans used hydraulic mining methods on a large scale to prospect for the veins of ore, especially a now obsolete form of mining known as hushing. This method involved building numerous aqueducts to supply water to the minehead where it was stored in large reservoirs and tanks. When a full tank was opened, the flood of water sluiced away the overburden to expose the bedrock underneath and any gold veins. The rock was then worked upon by fire-setting to heat the rock, which would be quenched with a stream of water. The resulting thermal shock cracked the rock, enabling it to be removed, aided by further streams of water from the overhead tanks. The Roman miners used similar methods to work cassiterite deposits in Cornwall and lead ore in the Pennines.

The methods had been developed by the Romans in Spain in 25 AD to exploit large alluvial gold deposits, the largest site being at Las Medulas, where seven long aqueducts were built to tap local rivers and to sluice the deposits. Spain was one of the most important mining regions, but all regions of the Roman Empire were exploited. In Great Britain the natives had mined minerals for millennia, but after the Roman conquest, the scale of the operations increased dramatically, as the Romans needed Britannia's resources, especially gold, silver, tin, and lead.

Roman techniques were not limited to surface mining. They followed the ore veins underground once opencast mining was no longer feasible. At Dolaucothi they stoped out the veins, and drove adits through barren rock to drain the stopes. The same adits were also used to ventilate the workings, especially important when fire-setting was used. At other parts of the site, they penetrated the water table and dewatered the mines using several kinds of machines, especially reverse overshot water-wheels. These were used extensively in the copper mines at Rio Tinto in Spain, where one sequence comprised 16 such wheels arranged in pairs, and lifting water about 80 feet (24 m). They were worked as treadmills with miners standing on the top slats. Many examples of such devices have been found in old Roman mines and some examples are now preserved in the British Museum and the National Museum of Wales.

Medieval Europe

Gallery, 12th to 13th century, Germany.

Mining as an industry underwent dramatic changes in medieval Europe. The mining industry in the early Middle Ages was mainly focused on the extraction of copper and iron. Other precious metals were also used, mainly for gilding or coinage. Initially, many metals were obtained through open-pit mining, and ore was primarily extracted from shallow depths, rather than through deep mine shafts. Around the 14th century, the growing use of weapons, armour, stirrups, and horseshoes greatly increased the demand for iron. Medieval knights, for example, were often laden with up to 100 pounds of plate or chain link armour in addition to swords, lances and other weapons. The overwhelming dependency on iron for military purposes spurred iron production and extraction processes.

The silver crisis of 1465 occurred when all mines had reached depths at which the shafts could no longer be pumped dry with the available technology. Although an increased use of bank notes, credit and copper coins during this period did decrease the value of, and dependence on, precious metals, gold and silver still remained vital to the story of medieval mining.

Due to differences in the social structure of society, the increasing extraction of mineral deposits spread from central Europe to England in the mid-sixteenth century. On the continent, all mineral deposits belonged to the crown, and this regalian right was stoutly maintained; but in England, it was pared down to gold and silver (of which there were virtually no deposits) by a judicial decision of 1568 and a law of 1688. England had iron, zinc, copper, lead, and tin ores. Landlords who owned the base metals and coal under their estates were now rendered with a strong inducement to extract these metals or to lease the deposits and collect royalties from mine operators. English,

German, and Dutch capital combined to finance extraction and refining. Hundreds of German technicians and skilled workers were brought over; in 1642 a colony of 4,000 foreigners was mining and smelting copper at Keswick in the northwestern mountains.

Use of water power in the form of water mills was extensive. The water mills were employed in crushing ore, raising ore from shafts, and ventilating galleries by powering giant bellows. Black powder was first used in mining in Selmecbánya, Kingdom of Hungary (now Banská Štiavnica, Slovakia) in 1627. Black powder allowed blasting of rock and earth to loosen and reveal ore veins. Blasting was much faster than fire-setting and allowed the mining of previously impenetrable metals and ores. In 1762, the world's first mining academy was established in the same town.

The widespread adoption of agricultural innovations such as the iron plowshare, as well as the growing use of metal as a building material, was also a driving force in the tremendous growth of the iron industry during this period. Inventions like the arrastra were often used by the Spanish to pulverize ore after being mined. This device was powered by animals and used the same principles used for grain threshing.

Much of the knowledge of medieval mining techniques comes from books such as Biringuccio's *De la pirotechnia* and probably most importantly from Georg Agricola's *De re metallica* (1556). These books detail many different mining methods used in German and Saxon mines. One of the prime issues confronting medieval miners (and one which Agricola explains in detail) was the removal of water from mining shafts. As miners dug deeper to access new veins, flooding became a very real obstacle. The mining industry became dramatically more efficient and prosperous with the invention of mechanical and animal driven pumps.

Classical Philippine Civilization

The image of a Maharlika class of the Philippine Society , depicted in Boxer Codex that the Gold used as a form of Jewelry (ca.1400).

Mining in the Philippines began around 1000 BC. The early Filipinos worked various mines of gold, silver, copper and iron. Jewels, gold ingots, chains, calombigas and earrings were handed down from antiquity and inherited from their ancestors. Gold dagger handles, gold dishes, tooth plating, and huge gold ornamets were also used. In Laszlo Legeza's "Tantric elements in pre-Hispanic Philippines Gold Art", he mentioned that gold jewelry of Philippine origin was found in Ancient Egypt. According to Antonio Pigafetta, the people of Mindoro possessed great skill in mixing gold with other metals and gave it a natural and perfect appearance that could deceive even the best of silversmiths. The natives were also known for the jewelries made of other precious stones such as carnelian, agate and pearl. Some outstanding examples of Philippine jewelry included necklaces, belts, armlets and rings placed around the waist.

The Americas

Lead mining in the upper Mississippi River region of the U.S., 1865.

There are ancient, prehistoric copper mines along Lake Superior, and metallic copper was still found there, near the surface, in colonial times. Indegenous peoples availed themselves of this copper starting at least 5,000 years ago," and copper tools, arrowheads, and other artifacts that were part of an extensive native trade network have been discovered. In addition, obsidian, flint, and other minerals were mined, worked, and traded. Early French explorers who encountered the sites[clarification needed] made no use of the metals due to the difficulties of transporting them, but the copper was eventually traded throughout the continent along major river routes.

Miners at the Tamarack Mine in Copper Country, Michigan, U.S. in 1905.

In the early colonial history of the Americas, "native gold and silver was quickly expropriated and sent back to Spain in fleets of gold- and silver-laden galleons," the gold and silver originating mostly from mines in Central and South America. Turquoise dated at 700 A.D. was mined in pre-Columbian America; in the Cerillos Mining District in New Mexico, estimates are that "about 15,000 tons of rock had been removed from Mt. Chalchihuitl using stone tools before 1700."

Mining in the United States became prevalent in the 19th century, and the General Mining Act of 1872 was passed to encourage mining of federal lands. As with the California Gold Rush in the mid-19th century, mining for minerals and precious metals, along with ranching, was a driving factor in the Westward Expansion to the Pacific coast. With the exploration of the West, mining camps were established and "expressed a distinctive spirit, an enduring legacy to the new nation;" Gold Rushers would experience the same problems as the Land Rushers of the transient West that preceded them. Aided by railroads, many traveled West for work opportunities in mining. Western cities such as Denver and Sacramento originated as mining towns.

When new areas were explored, it was usually the gold (placer and then load) and then silver that were taken into possession and extracted first. Other metals would often wait for railroads or canals, as coarse gold dust and nuggets do not require smelting and are easy to identify and transport.

Modern Period

In the early 20th century, the gold and silver rush to the western United States also stimulated mining for coal as well as base metals such as copper, lead, and iron. Areas in modern Montana, Utah, Arizona, and later Alaska became predominate suppliers of copper to the world, which was increasingly demanding copper for electrical and households goods. Canada's mining industry grew more slowly than did the United States' due to limitations in transportation, capital, and U.S. competition; Ontario was the major producer of the early 20th century with nickel, copper, and gold.

Meanwhile, Australia experienced the Australian gold rushes and by the 1850s was producing 40% of the world's gold, followed by the establishment of large mines such as the Mount Morgan Mine, which ran for nearly a hundred years, Broken Hill ore deposit (one of the largest zinc-lead ore deposits), and the iron ore mines at Iron Knob. After declines in production, another boom in mining occurred in the 1960s. Now, in the early 21st century, Australia remains a major world mineral producer.

As the 21st century begins, a globalized mining industry of large multinational corporations has arisen. Peak minerals and environmental impacts have also become a concern. Different elements, particularly rare earth minerals, have begun to increase in demand as a result of new technologies.

Mine Development and Lifecycle

The process of mining from discovery of an ore body through extraction of minerals and finally to returning the land to its natural state consists of several distinct steps. The first is discovery of the ore body, which is carried out through prospecting or exploration to find and then define the extent, location and value of the ore body. This leads to a mathematical resource estimation to estimate the size and grade of the deposit.

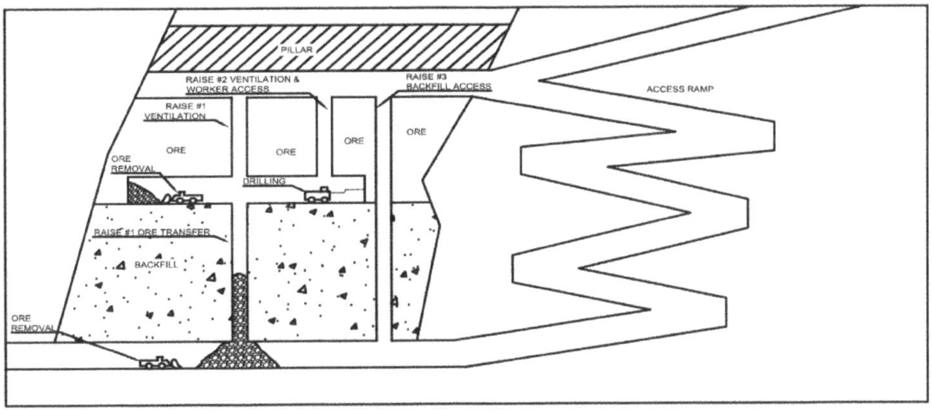

Schematic of a cut and fill mining operation in hard rock.

This estimation is used to conduct a pre-feasibility study to determine the theoretical economics of the ore deposit. This identifies, early on, whether further investment in estimation and engineering studies is warranted and identifies key risks and areas for further work. The next step is to conduct a feasibility study to evaluate the financial viability, the technical and financial risks, and the robustness of the project.

This is when the mining company makes the decision whether to develop the mine or to walk away from the project. This includes mine planning to evaluate the economically recoverable portion of the deposit, the metallurgy and ore recoverability, marketability and payability of the ore concentrates, engineering concerns, milling and infrastructure costs, finance and equity requirements, and an analysis of the proposed mine from the initial excavation all the way through to reclamation. The proportion of a deposit that is economically recoverable is dependent on the enrichment factor of the ore in the area.

To gain access to the mineral deposit within an area it is often necessary to mine through or remove waste material which is not of immediate interest to the miner. The total movement of ore and waste constitutes the mining process. Often more waste than ore is mined during the life of a mine, depending on the nature and location of the ore body. Waste removal and placement is a major cost to the mining operator, so a detailed characterization of the waste material forms an essential part of the geological exploration program for a mining operation.

Once the analysis determines a given ore body is worth recovering, development begins to create access to the ore body. The mine buildings and processing plants are built, and any necessary equipment is obtained. The operation of the mine to recover the ore begins and continues as long as the company operating the mine finds it economical to do so. Once all the ore that the mine can produce profitably is recovered, reclamation begins to make the land used by the mine suitable for future use.

Mining Techniques

Mining techniques can be divided into two common excavation types: surface mining and sub-surface (underground) mining. Today, surface mining is much more common, and produces, for example, 85% of minerals (excluding petroleum and natural gas) in the United States, including 98% of metallic ores.

Underground longwall mining.

Targets are divided into two general categories of materials: *placer deposits*, consisting of valuable minerals contained within river gravels, beach sands, and other unconsolidated materials; and *lode deposits*, where valuable minerals are found in veins, in layers, or in mineral grains generally distributed throughout a mass of actual rock. Both types of ore deposit, placer or lode, are mined by both surface and underground methods.

Some mining, including much of the rare earth elements and uranium mining, is done by less-common methods, such as in-situ leaching: this technique involves digging neither at the surface nor underground. The extraction of target minerals by this technique requires that they be soluble, e.g., potash, potassium chloride, sodium chloride, sodium sulfate, which dissolve in water. Some minerals, such as copper minerals and uranium oxide, require acid or carbonate solutions to dissolve.

Surface Mining

Surface mining is done by removing (stripping) surface vegetation, dirt, and, if necessary, layers of bedrock in order to reach buried ore deposits. Techniques of surface mining include: open-pit mining, which is the recovery of materials from an open pit in the ground, quarrying, identical to open-pit mining except that it refers to sand, stone and clay; strip mining, which consists of stripping surface layers off to reveal ore/seams underneath; and mountaintop removal, commonly associated with coal mining, which involves taking the top of a mountain off to reach ore deposits at depth. Most (but not all) placer deposits, because of their shallowly buried nature, are mined by surface methods. Finally, landfill mining involves sites where landfills are excavated and processed.

Garzweiler surface mine, Germany.

Underground Mining

Mantrip used for transporting miners within an underground mine.

Sub-surface mining consists of digging tunnels or shafts into the earth to reach buried ore deposits. Ore, for processing, and waste rock, for disposal, are brought to the surface through the tunnels and shafts. Sub-surface mining can be classified by the type of access shafts used, the extraction method or the technique used to reach the mineral deposit. Drift mining utilizes horizontal access tunnels, slope mining uses diagonally sloping access shafts, and shaft mining utilizes vertical access shafts. Mining in hard and soft rock formations require different techniques.

Other methods include shrinkage stope mining, which is mining upward, creating a sloping underground room, long wall mining, which is grinding a long ore surface underground, and room and pillar mining, which is removing ore from rooms while leaving pillars in place to support the roof of the room. Room and pillar mining often leads to retreat mining, in which supporting pillars are removed as miners retreat, allowing the room to cave in, thereby loosening more ore. Additional sub-surface mining methods include hard rock mining, which is mining of hard rock (igneous, metamorphic or sedimentary) materials, bore hole mining, drift and fill mining, long hole slope mining, sub level caving, and block caving.

Highwall Mining

Caterpillar Highwall Miner HW300 - Technology Bridging Underground and Open Pit Mining.

Highwall mining is another form of surface mining that evolved from auger mining. In Highwall mining, the coal seam is penetrated by a continuous miner propelled by a hydraulic Pushbeam Transfer Mechanism (PTM). A typical cycle includes sumping (launch-pushing forward) and shearing (raising and lowering the cutterhead boom to cut the entire height of the coal seam). As the coal recovery cycle continues, the cutterhead is progressively launched into the coal seam for 19.72 feet (6.01 m). Then, the Pushbeam Transfer Mechanism (PTM) automatically inserts a 19.72-foot (6.01 m) long rectangular Pushbeam (Screw-Conveyor Segment) into the center section of the machine between the Powerhead and the cutterhead. The Pushbeam system can penetrate nearly 1,000 feet (300 m) into the coal seam. One patented Highwall mining systems use augers enclosed inside the Pushbeam that prevent the mined coal from being contaminated by rock debris during the conveyance process. Using a video imaging and/or a gamma ray sensor and/or other Geo-Radar systems like a coal-rock interface detection sensor (CID), the operator can see ahead projection of the seam-rock interface and guide the continuous miner's progress. Highwall mining can produce thousands of tons of coal in contour-strip operations with narrow benches, previously mined areas, trench mine applications and steep-dip seams with controlled water-inflow pump system and/or a gas (inert) venting system.

Machines

Heavy machinery is used in mining to explore and develop sites, to remove and stockpile overburden, to break and remove rocks of various hardness and toughness, to process the ore, and to carry out reclamation projects after the mine is closed. Bulldozers, drills, explosives and trucks are all necessary for excavating the land. In the case of placer mining, unconsolidated gravel, or alluvium, is fed into machinery consisting of a hopper and a shaking screen or trommel which frees the desired minerals from the waste gravel. The minerals are then concentrated using sluices or jigs.

The Bagger 288 is a bucket-wheel excavator used in strip mining. It is also the largest land vehicle of all time.

A Bucyrus Erie 2570 dragline and CAT 797 haul truck at the North Antelope Rochelle opencut coal mine.

Large drills are used to sink shafts, excavate stopes, and obtain samples for analysis. Trams are used to transport miners, minerals and waste. Lifts carry miners into and out of mines, and move rock and ore out, and machinery in and out, of underground mines. Huge trucks, shovels and cranes are employed in surface mining to move large quantities of overburden and ore. Processing plants utilize large crushers, mills, reactors, roasters and other equipment to consolidate the mineral-rich material and extract the desired compounds and metals from the ore.

Processing

Once the mineral is extracted, it is often then processed. The science of extractive metallurgy is a specialized area in the science of metallurgy that studies the extraction of valuable metals from their ores, especially through chemical or mechanical means.

Mineral processing (or mineral dressing) is a specialized area in the science of metallurgy that studies the mechanical means of crushing, grinding, and washing that enable the separation (extractive metallurgy) of valuable metals or minerals from their gangue (waste material). Processing of placer ore material consists of gravity-dependent methods of separation, such as sluice boxes. Only minor shaking or washing may be necessary to disaggregate (unclump) the sands or gravels before processing. Processing of ore from a lode mine, whether it is a surface or subsurface mine, requires that the rock ore be crushed and pulverized before extraction of the valuable minerals begins. After lode ore is crushed, recovery of the valuable minerals is done by one, or a combination of several, mechanical and chemical techniques.

Since most metals are present in ores as oxides or sulfides, the metal needs to be reduced to its metallic form. This can be accomplished through chemical means such as smelting or through electrolytic reduction, as in the case of aluminium. Geometallurgy combines the geologic sciences with extractive metallurgy and mining.

Environmental Effects

Iron hydroxide precipitate stains a stream receiving acid drainage from surface coal mining.

Environmental issues can include erosion, formation of sinkholes, loss of biodiversity, and contamination of soil, groundwater and surface water by chemicals from mining processes. In some cases, additional forest logging is done in the vicinity of mines to create space for the storage of the created debris and soil. Contamination resulting from leakage of chemicals can also affect the health of the local population if not properly controlled. Extreme examples of pollution from mining activities include coal fires, which can last for years or even decades, producing massive amounts of environmental damage.

Mining companies in most countries are required to follow stringent environmental and rehabilitation codes in order to minimize environmental impact and avoid impacting human health. These codes and regulations all require the common steps of environmental impact assessment, development of environmental management plans, mine closure planning (which must be done before the start of mining operations), and environmental monitoring during operation and after closure. However, in some areas, particularly in the developing world, government regulations may not be well enforced.

For major mining companies and any company seeking international financing, there are a number of other mechanisms to enforce good environmental standards. These generally relate to financing standards such as the Equator Principles, IFC environmental standards, and criteria for Socially responsible investing. Mining companies have used this oversight from the financial sector to argue for some level of industry self-regulation. In 1992, a Draft Code of Conduct for Transnational Corporations was proposed at the Rio Earth Summit by the UN Centre for Transnational Corporations (UNCTC), but the Business Council for Sustainable Development (BCSD) together with the International Chamber of Commerce (ICC) argued successfully for self-regulation instead.

This was followed by the Global Mining Initiative which was begun by nine of the largest metals and mining companies and which led to the formation of the International Council on Mining and Metals, whose purpose was to "act as a catalyst" in an effort to improve social and environmental performance in the mining and metals industry internationally. The mining industry has provided funding to various conservation groups, some of which have been working with conservation agendas that are at odds with an emerging acceptance of the rights of indigenous people – particularly the right to make land-use decisions.

Certification of mines with good practices occurs through the International Organization for Standardization (ISO). For example, ISO 9000 and ISO 14001, which certify an "auditable environmental management system", involve short inspections, although they have been accused of lacking rigor. Certification is also available through Ceres' Global Reporting Initiative, but these reports are voluntary and unverified. Miscellaneous other certification programs exist for various projects, typically through nonprofit groups.

The purpose of a 2012 EPS PEAKS paper was to provide evidence on policies managing ecological costs and maximise socio-economic benefits of mining using host country regulatory initiatives. It found existing literature suggesting donors encourage developing countries to:

- Make the environment-poverty link and introduce cutting-edge wealth measures and natural capital accounts.
- Reform old taxes in line with more recent financial innovation, engage directly with the

companies, enacting land use and impact assessments, and incorporate specialised support and standards agencies.

- Set in play transparency and community participation initiatives using the wealth accrued.

Waste

Ore mills generate large amounts of waste, called tailings. For example, 99 tons of waste are generated per ton of copper, with even higher ratios in gold mining - because only 5.3 g of gold is extracted per ton of ore, a ton of gold produces 200,000 tons of tailings. These tailings can be toxic. Tailings, which are usually produced as a slurry, are most commonly dumped into ponds made from naturally existing valleys. These ponds are secured by impoundments (dams or embankment dams). In 2000 it was estimated that 3,500 tailings impoundments existed, and that every year, 2 to 5 major failures and 35 minor failures occurred; for example, in the Marcopper mining disaster at least 2 million tons of tailings were released into a local river. Subaqueous tailings disposal is another option. The mining industry has argued that submarine tailings disposal (STD), which disposes of tailings in the sea, is ideal because it avoids the risks of tailings ponds; although the practice is illegal in the United States and Canada, it is used in the developing world.

The waste is classified as either sterile or mineralised, with acid generating potential, and the movement and storage of this material forms a major part of the mine planning process. When the mineralised package is determined by an economic cut-off, the near-grade mineralised waste is usually dumped separately with view to later treatment should market conditions change and it becomes economically viable. Civil engineering design parameters are used in the design of the waste dumps, and special conditions apply to high-rainfall areas and to seismically active areas. Waste dump designs must meet all regulatory requirements of the country in whose jurisdiction the mine is located. It is also common practice to rehabilitate dumps to an internationally acceptable standard, which in some cases means that higher standards than the local regulatory standard are applied.

Renewable Energy and Mining

Many mining sites are remote and not connected to the grid. Electricity is typically generated with diesel generators. Due to high transportation cost and theft during transportation the cost for generating electricity is normally high. Renewable energy applications are becoming an alternative or amendment. Both solar and wind power plants can contribute in saving diesel costs at mining sites. Renewable energy applications have been built at mining sites. Cost savings can reach up to 70%.

Mining Industry

Mining exists in many countries. London is known as the capital of global "mining houses" such as Rio Tinto Group, BHP Billiton, and Anglo American PLC. The US mining industry is also large, but it is dominated by the coal and other nonmetal minerals (e.g., rock and sand), and various regulations have worked to reduce the significance of mining in the United States. In 2007 the total market capitalization of mining companies was reported at US$962 billion, which compares to a total global market cap of publicly traded companies of about US$50 trillion in 2007. In 2002, Chile and Peru were reportedly the major mining countries of South America. The mineral industry of Africa includes the mining of various minerals; it produces relatively little of the industrial metals copper, lead, and zinc, but according to one estimate has as a percent of world reserves 40% of gold, 60% of cobalt, and 90% of the world's platinum group metals. Mining in India is a significant part of that country's economy. In the developed world, mining in Australia, with BHP Billiton founded and headquartered in the country, and mining in Canada are particularly significant. For rare earth minerals mining, China reportedly controlled 95% of production in 2013.

The Bingham Canyon Mine of Rio Tinto's subsidiary, Kennecott Utah Copper.

While exploration and mining can be conducted by individual entrepreneurs or small businesses, most modern-day mines are large enterprises requiring large amounts of capital to establish. Consequently, the mining sector of the industry is dominated by large, often multinational, companies, most of them publicly listed. It can be argued that what is referred to as the 'mining industry' is actually two sectors, one specializing in exploration for new resources and the other in mining those resources. The exploration sector is typically made up of individuals and small mineral resource companies, called "juniors", which are dependent on venture capital. The mining sector is made up of large multinational companies that are sustained by production from their mining operations. Various other industries such as equipment manufacture, environmental testing, and metallurgy analysis rely on, and support, the mining industry throughout the world. Canadian stock exchanges have a particular focus on mining companies, particularly junior exploration companies through Toronto's TSX Venture Exchange; Canadian companies raise capital on these exchanges and then invest the money in exploration globally. Some have argued that below juniors there exists a substantial sector of illegitimate companies primarily focused on manipulating stock prices.

Mining operations can be grouped into five major categories in terms of their respective resources. These are oil and gas extraction, coal mining, metal ore mining, nonmetallic mineral mining and quarrying, and mining support activities. Of all of these categories, oil and gas extraction remains one of the largest in terms of its global economic importance. Prospecting potential mining sites, a vital area of concern for the mining industry, is now done using sophisticated new technologies such as seismic prospecting and remote-sensing satellites. Mining is heavily affected by the prices of the commodity minerals, which are often volatile. The 2000s commodities boom ("commodities supercycle") increased the prices of commodities, driving aggressive mining. In addition, the price of gold increased dramatically in the 2000s, which increased gold mining; for example, one study found that conversion of forest in the Amazon increased six-fold from the period 2003–2006 (292 ha/yr) to the period 2006–2009 (1,915 ha/yr), largely due to artisanal mining.

Corporate Classifications

Mining companies can be classified based on their size and financial capabilities:

- Major companies are considered to have an adjusted annual mining-related revenue of more than US$500 million, with the financial capability to develop a major mine on its own.
- Intermediate companies have at least $50 million in annual revenue but less than $500 million.
- Junior companies rely on equity financing as their principal means of funding exploration. Juniors are mainly pure exploration companies, but may also produce minimally, and do not have a revenue exceeding US$50 million.

Regulation and Governance

New regulation and process of legislative reforms aims to enrich the harmonization and stability of the mining sector in mineral-rich countries. The new legislation for mining industry in the African countries still appears as an emerging issue with a potential to be solved, until a consensus is reached on the best approach. By the beginning of the 21st century the booming and more complex mining sector in mineral-rich countries provided only slight benefits to local communities in terms of sustainability. Increasing debates and influence by NGOs and communities appealed for a new program which would have had also included a disadvantaged communities, and would have had worked towards sustainable development even after mine closure (included transparency and revenue management). By the early 2000s, community development issues and resettlements became mainstreamed in World Bank mining projects. Mining-industry expansion after an increase of mineral prices in 2003 and also potential fiscal revenues in those countries created an omission in the other economic sectors in terms of finances and development. Furthermore, it had highlighted regional and local demand of mining-revenues and lack of ability of sub-national governments to use the revenues. The Fraser Institute (a Canadian think tank) has highlighted the environmental protection laws in developing countries, as well as the voluntary efforts by mining companies to improve their environmental impact.

In 2007 the Extractive Industries Transparency Initiative (EITI) was mainstreamed in all countries cooperating with the World Bank in mining industry reform. The EITI is operating and implementing with a support of EITI Multi-Donor Trust Fund, managed by The World Bank. The Extractive Industries Transparency Initiative (EITI) aims to increase transparency in transactions between governments and companies within extractive industries by monitoring the revenues and benefits between industries and recipient governments. The entrance process is voluntary for each country and is being monitored by multi-stakeholders involving government, private companies and civil society representatives, responsible for disclosure and dissemination of the reconciliation report; however, the competitive disadvantage of company-by company public report is for some of the businesses in Ghana, the main constraint. Therefore, the outcome assessment in terms of failure or success of the new EITI regulation does not only "rest on the government's shoulders" but also on civil society and companies.

On the other hand, criticism points out two main implementation issues; inclusion or exclusion of artisanal mining and small-scale mining (ASM) from the EITI and how to deal with "non-cash" payments made by companies to subnational governments. Furthermore, disproportion of the revenues mining industry creates to the comparatively small number of people that it employs, causes another controversy. The issue of artisanal mining is clearly an issue in EITI Countries such as the Central African Republic, D.R. Congo, Guinea, Liberia and Sierra Leone – i.e. almost half of the mining countries implementing the EITI. Among other things, limited scope of the EITI involving disparity in terms of knowledge of the industry and negotiation skills, thus far flexibility of the policy (e.g. liberty of the countries to expand beyond the minimum requirements and adapt it to their needs), creates another risk of unsuccessful implementation. Public awareness increase, where government should act as a bridge between public and initiative for a successful outcome of the policy is an important element to be considered.

World Bank

The World Bank has been involved in mining since 1955, mainly through grants from its International Bank for Reconstruction and Development, with the Bank's Multilateral Investment Guarantee Agency offering political risk insurance. Between 1955 and 1990 it provided about $2 billion to fifty mining projects, broadly categorized as reform and rehabilitation, greenfield mine construction, mineral processing, technical assistance, and engineering. These projects have been criticized, particularly the Ferro Carajas project of Brazil, begun in 1981. The World Bank established mining codes intended to increase foreign investment; in 1988 it solicited feedback from 45 mining companies on how to increase their involvement.

In 1992 the World Bank began to push for privatization of government-owned mining companies with a new set of codes, beginning with its report *The Strategy for African Mining*. In 1997, Latin America's largest miner Companhia Vale do Rio Doce (CVRD) was privatized. These and other developments such as the Philippines 1995 Mining Act led the bank to publish a third report (*Assistance for Minerals Sector Development and Reform in Member Countries*) which endorsed mandatory environment impact assessments and attention to the concerns of the local population. The codes based on this report are influential in the legislation of developing nations. The new codes are intended to encourage development through tax holidays, zero custom duties, reduced income taxes, and related measures. The results of these codes were analyzed by a group from the University of Quebec, which concluded that the codes promote foreign investment but "fall very short of permitting sustainable development". The observed negative correlation between natural resources and economic development is known as the resource curse.

Safety

Safety has long been a concern in the mining business, especially in sub-surface mining. The Courrières mine disaster, Europe's worst mining accident, involved the death of 1,099 miners in Northern France on March 10, 1906. This disaster was surpassed only by the Benxihu Colliery accident in China on April 26, 1942, which killed 1,549 miners. While mining today is substantially safer than it was in previous decades, mining accidents still occur. Government figures indicate that

5,000 Chinese miners die in accidents each year, while other reports have suggested a figure as high as 20,000. Mining accidents continue worldwide, including accidents causing dozens of fatalities at a time such as the 2007 Ulyanovskaya Mine disaster in Russia, the 2009 Heilongjiang mine explosion in China, and the 2010 Upper Big Branch Mine disaster in the United States.

There are numerous occupational hazards associated with mining, including exposure to rockdust which can lead to diseases such as silicosis, asbestosis, and pneumoconiosis. Gases in the mine can lead to asphyxiation and could also be ignited. Mining equipment can generate considerable noise, putting workers at risk for hearing loss. Cave-ins, rock falls, and exposure to excess heat are also known hazards.

Proper ventilation, hearing protection, and spraying equipment with water are important safety practices in mines.

Records

As of 2008, the deepest mine in the world is TauTona in Carletonville, South Africa at 3.9 kilometres (2.4 mi), replacing the neighboring Savuka Mine in the North West Province of South Africa at 3,774 metres (12,382 ft). East Rand Mine in Boksburg, South Africa briefly held the record at 3,585 metres (11,762 ft), and the first mine declared the deepest in the world was also TauTona when it was at 3,581 metres (11,749 ft).

The Moab Khutsong gold mine in North West Province (South Africa) has the world's longest winding steel wire rope, able to lower workers to 3,054 metres (10,020 ft) in one uninterrupted four-minute journey.

The deepest mine in Europe is the 16th shaft of the uranium mines in Příbram, Czech Republic at 1,838 metres (6,030 ft), second is Bergwerk Saar in Saarland, Germany at 1,750 metres (5,740 ft).

The deepest open-pit mine in the world is Bingham Canyon Mine in Bingham Canyon, Utah, United States at over 1,200 metres (3,900 ft). The largest and second deepest open-pit copper mine in the world is Chuquicamata in Chuquicamata, Chile at 900 metres (3,000 ft), 443,000 tons of copper and 20,000 tons of molybdenum produced annually.

The deepest open-pit mine with respect to sea level is Tagebau Hambach in Germany, where the base of the pit is 293 metres (961 ft) below sea level.

The largest underground mine is Kiirunavaara Mine in Kiruna, Sweden. With 450 kilometres (280 mi) of roads, 40 million tonnes of ore produced yearly, and a depth of 1,270 metres (4,170 ft), it is also one of the most modern underground mines. The deepest borehole in the world is Kola Superdeep Borehole at 12,262 metres (40,230 ft). This, however, is not a matter of mining but rather related to scientific drilling.

Metal Reserves and Recycling

During the twentieth century, the variety of metals used in society grew rapidly. Today, the development of major nations such as China and India and advances in technologies are fueling an ever greater demand. The result is that metal mining activities are expanding and more and more of the world's metal stocks are above ground in use rather than below ground as unused reserves. An example is the in-use stock of copper. Between 1932 and 1999, copper in use in the USA rose from 73 kilograms (161 lb) to 238 kilograms (525 lb) per person.

95% of the energy used to make aluminium from bauxite ore is saved by using recycled material. However, levels of metals recycling are generally low. In 2010, the International Resource Panel, hosted by the United Nations Environment Programme (UNEP), published reports on metal stocks that exist within society and their recycling rates.

The report's authors observed that the metal stocks in society can serve as huge mines above ground. However, they warned that the recycling rates of some rare metals used in applications such as mobile phones, battery packs for hybrid cars, and fuel cells are so low that unless future end-of-life recycling rates are dramatically stepped up these critical metals will become unavailable for use in modern technology.

As recycling rates are low and so much metal has already been extracted, some landfills now contain higher concentrations of metal than mines themselves. This is especially true with aluminium, found in cans, and precious metals in discarded electronics. Furthermore, waste after 15 years has still not broken down, so less processing would be required when compared to mining ores. A study undertaken by Cranfield University has found £360 million of metals could be mined from just 4 landfill sites. There is also up to 20MW/kg of energy in waste, potentially making the re-extraction more profitable. However, although the first landfill mine opened in Tel Aviv, Israel in 1953, little work has followed due to the abundance of accessible ores.

Casting

Casting is a manufacturing process in which a liquid material is usually poured into a mold, which contains a hollow cavity of the desired shape, and then allowed to solidify. The solidified part is also known as a casting, which is ejected or broken out of the mold to complete the process. Casting materials are usually metals or various cold setting materials that cure after mixing two or more components together; examples are epoxy, concrete, plaster and clay. Casting is most often used for making complex shapes that would be otherwise difficult or uneconomical to make by other methods.

Casting is a 6000-year-old process. The oldest surviving casting is a copper frog from 3200 BC.

Molten metal before casting.

Casting iron in a sand mold.

Judenplatz Holocaust Memorial (Nameless Library), by Rachel Whiteread. Concrete cast of books on library shelves turned inside out.

Types

Metal

In metalworking, metal is heated until it becomes liquid and is then poured into a mold. The mold is a hollow cavity that includes the desired shape, but the mold also includes runners and vents that enable the metal to fill the mold. The mold and the metal are then cooled until the metal solidifies. The solidified part (the casting) is then recovered from the mold. Subsequent operations remove excess material caused by the casting process (e.g., such as the runners and vents).

Plaster, Concrete, or Plastic Resin

Plaster and other chemical curing materials such as concrete and plastic resin may be cast using single-use *waste* molds as noted above, multiple-use 'piece' molds, or molds made of small rigid pieces or of flexible material such as latex rubber (which is in turn supported by an exterior mold). When casting plaster or concrete, the material surface is flat and lacks transparency. Often topical treatments are applied to the surface. For example, painting and etching can be used in a way that give the appearance of metal or stone. Alternatively, the material is altered in its initial casting process and may contain colored sand so as to give an appearance of stone. By casting concrete, rather than plaster, it is possible to create sculptures, fountains, or seating for outdoor use. A simulation of high-quality marble may be made using certain chemically-set plastic resins (for example epoxy or polyester) with powdered stone added for coloration, often with multiple colors worked in. The latter is a common means of making washstands, washstand tops and shower stalls, with the skilled working of multiple colors resulting in simulated staining patterns as is often found in natural marble or travertine.

Fettling

Raw castings often contain irregularities caused by seams and imperfections in the molds, as well as access ports for pouring material into the molds. The process of cutting, grinding, shaving or sanding away these unwanted bits is called "fettling". In modern times robotic processes have been developed to perform some of the more repetitive parts of the fettling process, but historically fettlers carried out this arduous work manually, and often in conditions dangerous to their health.

Fettling can add significantly to the cost of the resulting product, and designers of molds seek to minimize it through the shape of the mold, the material being cast, and sometimes by including decorative elements.

Casting Process Simulation

Casting process simulation uses numerical methods to calculate cast component quality considering mold filling, solidification and cooling, and provides a quantitative prediction of casting mechanical properties, thermal stresses and distortion. Simulation accurately describes a cast component's quality up-front before production starts. The casting rigging can be designed with respect to the required component properties. This has benefits beyond a reduction in pre-production sampling, as the precise layout of the complete casting system also leads to energy, material, and tooling savings.

The software supports the user in component design, the determination of melting practice and casting methoding through to pattern and mold making, heat treatment, and finishing. This saves costs along the entire casting manufacturing route.

Casting process simulation was initially developed at universities starting from the early '70s, mainly in Europe and in the U.S., and is regarded as the most important innovation in casting technology over the last 50 years. Since the late '80s, commercial programs (such as AutoCAST and MAGMA) are available which make it possible for foundries to gain new insight into what is happening inside the mold or die during the casting process.

Melting

Melting (also known as fusion) is a physical process that results in the phase transition of a substance from a solid to a liquid. This occurs when the internal energy of the solid increases, typically by the application of heat or pressure, which increases the substance's temperature to the melting point. At the melting point, the ordering of ions or molecules in the solid breaks down to a less ordered state, and the solid melts to become a liquid. An object that has melted completely is molten (although this word is typically used for substances that melt only at a high temperature, such as molten iron or molten lava).

Ice melting

Substances in the molten state generally have reduced viscosity as the temperature increases. An exception to this principle is the element sulfur, whose viscosity increases to a point due to polymerization and then decreases with higher temperatures in its molten state.

Some organic compounds melt through mesophases, states of partial order between solid and liquid.

Melting as a First-Order Phase Transition

From a thermodynamics point of view, at the melting point the change in Gibbs free energy ΔG of the material is zero, but there are non-zero changes in the enthalpy (H) and the entropy (S), known respectively as the enthalpy of fusion (or latent heat of fusion) and the entropy of fusion. Melting is therefore classified as a first-order phase transition. Melting occurs when the Gibbs free energy of the liquid becomes lower than the solid for that material. The temperature at which this occurs is dependent on the ambient pressure.

Low-temperature helium is the only known exception to the general rule. Helium-3 has a negative enthalpy of fusion at temperatures below 0.3 K. Helium-4 also has a very slightly negative enthalpy of fusion below 0.8 K. This means that, at appropriate constant pressures, heat must be *removed* from these substances in order to melt them.

Melting Criteria

Among the theoretical criteria for melting, the Lindemann and of Born criteria are those most frequently used as a basis to analyse the melting conditions . The Lindemann criterion states that melting occurs because of vibrational instability, e.g. crystals melt when the average amplitude of thermal vibrations of atoms is relatively high compared with interatomic distances, e.g. $<\delta u^2>^{1/2} > \delta_L R_s$, where δu is the atomic displacement, the Lindemann parameter $\delta_L \approx 0.20...0.25$ and R_s is one-half of the inter-atomic distance. The Lindemann melting criterion is supported by experimental data both for crystalline materials and for glass-liquid transitions in amorphous materials. The Born criterion is based on rigidity catastrophe caused by the vanishing elastic shear modulus, e.g. when the crystal no longer has sufficient rigidity to mechanically withstand load.

Supercooling

Under a standard set of conditions, the melting point of a substance is a characteristic property. The melting point is often equal to the freezing point. However, under carefully created conditions, supercooling or superheating past the melting or freezing point can occur. Water on a very clean glass surface will often supercool several degrees below the freezing point without freezing. Fine emulsions of pure water have been cooled to −38 degrees Celsius without nucleation to form ice.. Nucleation occurs due to fluctuations in the properties of the material. If the material is kept still there is often nothing (such a physical vibration) to trigger this change, and supercooling (or superheating) may occur. Thermodynamically, the supercooled liquid is in the metastable state with respect to the crystalline phase, and it is likely to crystallize suddenly.

Melting of Amorphous Solids (Glasses)

Glasses are amorphous solids which are usually fabricated when the molten material cools very rapidly to below its glass transition temperature, without sufficient time for a regular crystal lattice to form. Solids are characterised by a high degree of connectivity between their molecules, and fluids have lower connectivity of their structural blocks. Melting of a solid material can also be considered as a percolation via broken connections between particles e.g. connecting bonds. In this approach melting of an amorphous material occurs when the broken bonds form a percolation cluster with T_g dependent on quasi-equilibrium thermodynamic parameters of bonds e.g. on enthalpy (H_d) and entropy (S_d) of formation of bonds in a given system at given conditions:

$$T_g = \frac{H_d}{S_d + R\ln(\frac{1-f_c}{f_c})}$$

where f_c is the percolation threshold and R is the universal gas constant. Although H_d and S_d are not true equilibrium thermodynamic parameters and can depend on the cooling rate of a melt they can be found from available experimental data on viscosity of amorphous materials.

Even below its melting point, quasi-liquid films can be observed on crystalline surfaces. The thickness of the film is temperature dependent. This effect is common for all crystalline materials. Premelting shows its effects in e.g. frost heave, the growth of snowflakes and, taking grain boundary interfaces into account, maybe even in the movement of glaciers.

Related Concepts

In genetics, melting DNA means to separate the double-stranded DNA into two single strands by heating or the use of chemical agents, cf. polymerase chain reaction.

Welding

Gas metal arc welding (MIG welding)

Welding is a fabrication or sculptural process that joins materials, usually metals or thermoplastics, by causing fusion, which is distinct from lower temperature metal-joining techniques such as brazing and soldering, which do not melt the base metal. In addition to melting the base metal, a filler material is often added to the joint to form a pool of molten material (the weld pool) that cools to form a joint that can be as strong, or even stronger, than the base material. Pressure may also be used in conjunction with heat, or by itself, to produce a weld.

Although less common, there are also solid state welding processes such as friction welding or shielded active gas welding in which metal does not melt.

Some of the best known welding methods include:

- Shielded metal arc welding (SMAW) – also known as "stick welding or electric welding", uses an electrode that has flux around it to protect the weld puddle. The electrode holder holds the electrode as it slowly melts away. Slag protects the weld puddle from atmospheric contamination.
- Gas tungsten arc welding (GTAW) – also known as TIG (tungsten, inert gas), uses a non-consumable tungsten electrode to produce the weld. The weld area is protected from atmospheric contamination by an inert shielding gas such as argon or helium.
- Gas metal arc welding (GMAW) – commonly termed MIG (metal, inert gas), uses a wire feeding gun that feeds wire at an adjustable speed and flows an argon-based shielding gas or a mix of argon and carbon dioxide (CO_2) over the weld puddle to protect it from atmospheric contamination.
- Flux-cored arc welding (FCAW) – almost identical to MIG welding except it uses a special tubular wire filled with flux; it can be used with or without shielding gas, depending on the filler.
- Submerged arc welding (SAW) – uses an automatically fed consumable electrode and a blanket of granular fusible flux. The molten weld and the arc zone are protected from atmospheric contamination by being "submerged" under the flux blanket.
- Electroslag welding (ESW) – a highly productive, single pass welding process for thicker materials between 1 inch (25 mm) and 12 inches (300 mm) in a vertical or close to vertical position.

Many different energy sources can be used for welding, including a gas flame, an electric arc, a laser, an electron beam, friction, and ultrasound. While often an industrial process, welding may be performed in many different environments, including in open air, under water, and in outer space. Welding is a hazardous undertaking and precautions are required to avoid burns, electric shock, vision damage, inhalation of poisonous gases and fumes, and exposure to intense ultraviolet radiation.

Until the end of the 19th century, the only welding process was forge welding, which blacksmiths had used for centuries to join iron and steel by heating and hammering. Arc welding and oxyfuel welding were among the first processes to develop late in the century, and electric resistance welding followed soon after. Welding technology advanced quickly during the early 20th century as the world wars drove the demand for reliable and inexpensive joining methods. Following the wars, several modern welding techniques were developed, including manual methods like SMAW, now one of the most popular welding methods, as well as semi-automatic and automatic processes such as GMAW, SAW, FCAW and ESW. Developments continued with the invention of laser beam welding, electron beam welding, magnetic pulse welding (MPW), and friction stir welding in the latter half of the century. Today, the science continues to advance. Robot welding is commonplace in industrial settings, and researchers continue to develop new welding methods and gain greater understanding of weld quality.

History

The history of joining metals goes back several millennia. Called forge welding, the earliest examples come from the Bronze and Iron Ages in Europe and the Middle East. The ancient Greek historian Herodotus states in *The Histories* of the 5th century BC that Glaucus of Chios "was the man who single-handedly invented iron welding". Welding was used in the construction of the Iron pillar of Delhi, erected in Delhi, India about 310 AD and weighing 5.4 metric tons.

The Middle Ages brought advances in forge welding, in which blacksmiths pounded heated metal repeatedly until bonding occurred. In 1540, Vannoccio Biringuccio published *De la pirotechnia*, which includes descriptions of the forging operation. Renaissance craftsmen were skilled in the process, and the industry continued to grow during the following centuries.

In 1800, Sir Humphry Davy discovered the short-pulse electrical arc and presented his results in 1801. In 1802, Russian scientist Vasily Petrov created the continuous electric arc, and subsequently published "News of Galvanic-Voltaic Experiments" in 1803, in which he described experiments carried out in 1802. Of great importance in this work was the description of a stable arc discharge and the indication of its possible use for many applications, one being melting metals. In 1808, Davy, who was unaware of Petrov's work, rediscovered the continuous electric arc. In 1881–82 inventors Nikolai Benardos (Russian) and Stanisław Olszewski (Polish) created the first electric arc welding method known as carbon arc welding using carbon electrodes. The advances in arc welding continued with the invention of metal electrodes in the late 1800s by a Russian, Nikolai Slavyanov (1888), and an American, C. L. Coffin (1890). Around 1900, A. P. Strohmenger released a coated metal electrode in Britain, which gave a more stable arc. In 1905, Russian scientist Vladimir Mitkevich proposed using a three-phase electric arc for welding. In 1919, alternating current welding was invented by C. J. Holslag but did not become popular for another decade.

Resistance welding was also developed during the final decades of the 19th century, with the first patents going to Elihu Thomson in 1885, who produced further advances over the next 15 years. Thermite welding was invented in 1893, and around that time another process, oxyfuel welding, became well established. Acetylene was discovered in 1836 by Edmund Davy, but its use was not practical in welding until about 1900, when a suitable torch was developed. At first, oxyfuel welding was one of the more popular welding methods due to its portability and relatively low cost. As the 20th century progressed, however, it fell out of favor for industrial applications. It was largely replaced with arc welding, as metal coverings (known as flux) for the electrode that stabilize the arc and shield the base material from impurities continued to be developed.

Bridge of Maurzyce.

World War I caused a major surge in the use of welding processes, with the various military powers attempting to determine which of the several new welding processes would be best. The British primarily used arc welding, even constructing a ship, the "Fullagar" with an entirely welded hull. Arc welding was first applied to aircraft during the war as well, as some German airplane fuselages were constructed using the process. Also noteworthy is the first welded road bridge in the world, the Maurzyce Bridge designed by Stefan Bryła of the Lwów University of Technology in 1927, and built across the river Słudwia near Łowicz, Poland in 1928.

Acetylene welding on cylinder water jacket, 1918.

During the 1920s, major advances were made in welding technology, including the introduction of automatic welding in 1920, in which electrode wire was fed continuously. Shielding gas became a subject receiving much attention, as scientists attempted to protect welds from the effects of oxygen and nitrogen in the atmosphere. Porosity and brittleness were the primary problems, and the solutions that developed included the use of hydrogen, argon, and helium as welding atmospheres.

During the following decade, further advances allowed for the welding of reactive metals like aluminum and magnesium. This in conjunction with developments in automatic welding, alternating current, and fluxes fed a major expansion of arc welding during the 1930s and then during World War II. In 1930, the first all-welded merchant vessel, M/S Carolinian, was launched.

During the middle of the century, many new welding methods were invented. 1930 saw the release of stud welding, which soon became popular in shipbuilding and construction. Submerged arc welding was invented the same year and continues to be popular today. In 1932 a Russian, Konstantin Khrenov successfully implemented the first underwater electric arc welding. Gas tungsten arc welding, after decades of development, was finally perfected in 1941, and gas metal arc welding followed in 1948, allowing for fast welding of non-ferrous materials but requiring expensive shielding gases. Shielded metal arc welding was developed during the 1950s, using a flux-coated consumable electrode, and it quickly became the most popular metal arc welding process. In 1957, the flux-cored arc welding process debuted, in which the self-shielded wire electrode could be used with automatic equipment, resulting in greatly increased welding speeds, and that same year, plasma arc welding was invented. Electroslag welding was introduced in 1958, and it was followed by its cousin, electrogas welding, in 1961. In 1953 the Soviet scientist N. F. Kazakov proposed the diffusion bonding method.

Other recent developments in welding include the 1958 breakthrough of electron beam welding, making deep and narrow welding possible through the concentrated heat source. Following the invention of the laser in 1960, laser beam welding debuted several decades later, and has proved to be especially useful in high-speed, automated welding. Magnetic pulse welding (MPW) is industrially used since 1967. Friction stir welding was invented in 1991 by Wayne Thomas at The Welding Institute (TWI, UK) and found high-quality applications all over the world. All of these four new processes continue to be quite expensive due the high cost of the necessary equipment, and this has limited their applications.

Processes

Arc

Man welding a metal structure in a newly constructed house in Bengaluru, India.

These processes use a welding power supply to create and maintain an electric arc between an electrode and the base material to melt metals at the welding point. They can use either direct (DC) or alternating (AC) current, and consumable or non-consumable electrodes. The welding region is sometimes protected by some type of inert or semi-inert gas, known as a shielding gas, and filler material is sometimes used as well.

Power Supplies

To supply the electrical power necessary for arc welding processes, a variety of different power supplies can be used. The most common welding power supplies are constant current power supplies and constant voltage power supplies. In arc welding, the length of the arc is directly related to the voltage, and the amount of heat input is related to the current. Constant current power supplies are most often used for manual welding processes such as gas tungsten arc welding and shielded metal arc welding, because they maintain a relatively constant current even as the voltage varies. This is important because in manual welding, it can be difficult to hold the electrode perfectly steady, and as a result, the arc length and thus voltage tend to fluctuate. Constant voltage power supplies hold the voltage constant and vary the current, and as a result, are most often used for automated welding processes such as gas metal arc welding, flux cored arc welding, and submerged arc welding. In these processes, arc length is kept constant, since any fluctuation in the distance between the wire and the base material is quickly rectified by a large change in current. For example, if the wire and the base material get too close, the current will rapidly increase, which in turn causes the heat to increase and the tip of the wire to melt, returning it to its original separation distance.

The type of current used plays an important role in arc welding. Consumable electrode processes such as shielded metal arc welding and gas metal arc welding generally use direct current, but the electrode can be charged either positively or negatively. In welding, the positively charged anode will have a greater heat concentration, and as a result, changing the polarity of the electrode affects weld properties. If the electrode is positively charged, the base metal will be hotter, increasing weld penetration and welding speed. Alternatively, a negatively charged electrode results in more shallow welds. Nonconsumable electrode processes, such as gas tungsten arc welding, can use either type of direct current, as well as alternating current. However, with direct current, because the electrode only creates the arc and does not provide filler material, a positively charged electrode causes shallow welds, while a negatively charged electrode makes deeper welds. Alternating current rapidly moves between these two, resulting in medium-penetration welds. One disadvantage of AC, the fact that the arc must be re-ignited after every zero crossing, has been addressed with the invention of special power units that produce a square wave pattern instead of the normal sine wave, making rapid zero crossings possible and minimizing the effects of the problem.

Processes

One of the most common types of arc welding is shielded metal arc welding (SMAW); it is also known as manual metal arc welding (MMA) or stick welding. Electric current is used to strike an arc between the base material and consumable electrode rod, which is made of filler material (typically steel) and is covered with a flux that protects the weld area from oxidation and contamination by producing carbon dioxide (CO_2) gas during the welding process. The electrode core itself acts as filler material, making a separate filler unnecessary.

Shielded metal arc welding.

The process is versatile and can be performed with relatively inexpensive equipment, making it well suited to shop jobs and field work. An operator can become reasonably proficient with a modest amount of training and can achieve mastery with experience. Weld times are rather slow, since the consumable electrodes must be frequently replaced and because slag, the residue from the flux, must be chipped away after welding. Furthermore, the process is generally limited to welding ferrous materials, though special electrodes have made possible the welding of cast iron, nickel, aluminum, copper, and other metals.

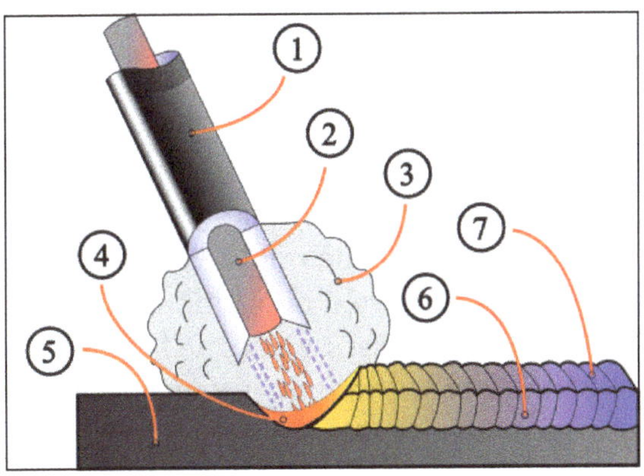

Diagram of arc and weld area, in shielded metal arc welding.

1. Coating Flow
2. Rod
3. Shield Gas
4. Fusion
5. Base metal
6. Weld metal
7. Solidified Slag

Gas metal arc welding (GMAW), also known as metal inert gas or MIG welding, is a semi-automatic or automatic process that uses a continuous wire feed as an electrode and an inert or semi-inert gas mixture to protect the weld from contamination. Since the electrode is continuous, welding speeds are greater for GMAW than for SMAW.

A related process, flux-cored arc welding (FCAW), uses similar equipment but uses wire consisting of a steel electrode surrounding a powder fill material. This cored wire is more expensive than the standard solid wire and can generate fumes and/or slag, but it permits even higher welding speed and greater metal penetration.

Gas tungsten arc welding (GTAW), or tungsten inert gas (TIG) welding, is a manual welding process that uses a nonconsumable tungsten electrode, an inert or semi-inert gas mixture, and a separate filler material. Especially useful for welding thin materials, this method is characterized by a stable arc and high quality welds, but it requires significant operator skill and can only be accomplished at relatively low speeds.

GTAW can be used on nearly all weldable metals, though it is most often applied to stainless steel and light metals. It is often used when quality welds are extremely important, such as in bicycle, aircraft and naval applications. A related process, plasma arc welding, also uses a tungsten electrode but uses plasma gas to make the arc. The arc is more concentrated than the GTAW arc, making transverse control more critical and thus generally restricting the technique to a mechanized process. Because of its stable current, the method can be used on a wider range of material thicknesses than can the GTAW process and it is much faster. It can be applied to all of the same materials as GTAW except magnesium, and automated welding of stainless steel is one important application of the process. A variation of the process is plasma cutting, an efficient steel cutting process.

Submerged arc welding (SAW) is a high-productivity welding method in which the arc is struck beneath a covering layer of flux. This increases arc quality, since contaminants in the atmosphere are blocked by the flux. The slag that forms on the weld generally comes off by itself, and combined with the use of a continuous wire feed, the weld deposition rate is high. Working conditions are much improved over other arc welding processes, since the flux hides the arc and almost no smoke is produced. The process is commonly used in industry, especially for large products and in the manufacture of welded pressure vessels. Other arc welding processes include atomic hydrogen welding, electroslag welding, electrogas welding, and stud arc welding.

Gas Welding

The most common gas welding process is oxyfuel welding, also known as oxyacetylene welding. It is one of the oldest and most versatile welding processes, but in recent years it has become less popular in industrial applications. It is still widely used for welding pipes and tubes, as well as repair work.

The equipment is relatively inexpensive and simple, generally employing the combustion of acetylene in oxygen to produce a welding flame temperature of about 3100 °C. The flame, since it is less concentrated than an electric arc, causes slower weld cooling, which can lead to greater residual stresses and weld distortion, though it eases the welding of high alloy steels. A similar process, generally called oxyfuel cutting, is used to cut metals.

Resistance

Resistance welding involves the generation of heat by passing current through the resistance caused by the contact between two or more metal surfaces. Small pools of molten metal are formed at the weld area as high current (1000–100,000 A) is passed through the metal. In general, resistance welding methods are efficient and cause little pollution, but their applications are somewhat limited and the equipment cost can be high.

Spot welder.

Spot welding is a popular resistance welding method used to join overlapping metal sheets of up to 3 mm thick. Two electrodes are simultaneously used to clamp the metal sheets together and to pass current through the sheets. The advantages of the method include efficient energy use, limited workpiece deformation, high production rates, easy automation, and no required filler materials. Weld strength is significantly lower than with other welding methods, making the process suitable for only certain applications. It is used extensively in the automotive industry—ordinary cars can have several thousand spot welds made by industrial robots. A specialized process, called shot welding, can be used to spot weld stainless steel.

Like spot welding, seam welding relies on two electrodes to apply pressure and current to join metal sheets. However, instead of pointed electrodes, wheel-shaped electrodes roll along and often feed the workpiece, making it possible to make long continuous welds. In the past, this process was used in the manufacture of beverage cans, but now its uses are more limited. Other resistance welding methods include butt welding, flash welding, projection welding, and upset welding.

Energy Beam

Energy beam welding methods, namely laser beam welding and electron beam welding, are relatively new processes that have become quite popular in high production applications. The two processes are quite similar, differing most notably in their source of power. Laser beam welding employs a highly focused laser beam, while electron beam welding is done in a vacuum and uses an electron beam. Both have a very high energy density, making deep weld penetration possible and minimizing the size of the weld area. Both processes are extremely fast, and are easily automated, making them highly productive. The primary disadvantages are their very high equipment costs (though these are decreasing) and a susceptibility to thermal cracking. Developments in this area include laser-hybrid welding, which uses principles from both laser beam welding and arc welding for even better weld properties, laser cladding, and x-ray welding.

Solid-state

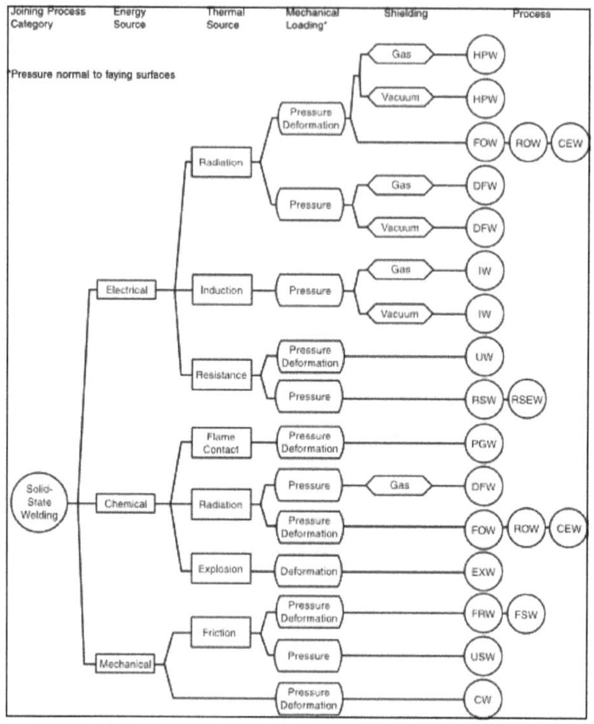

Solid-state welding processes classification chart.

Like the first welding process, forge welding, some modern welding methods do not involve the melting of the materials being joined. One of the most popular, ultrasonic welding, is used to connect thin sheets or wires made of metal or thermoplastic by vibrating them at high frequency and under high pressure. The equipment and methods involved are similar to that of resistance welding, but instead of electric current, vibration provides energy input. Welding metals with this process does not involve melting the materials; instead, the weld is formed by introducing mechanical vibrations horizontally under pressure. When welding plastics, the materials should have similar melting temperatures, and the vibrations are introduced vertically. Ultrasonic welding is commonly used for making electrical connections out of aluminum or copper, and it is also a very common polymer welding process.

Another common process, explosion welding, involves the joining of materials by pushing them together under extremely high pressure. The energy from the impact plasticizes the materials, forming a weld, even though only a limited amount of heat is generated. The process is commonly used for welding dissimilar materials, such as the welding of aluminum with steel in ship hulls or compound plates. Other solid-state welding processes include friction welding (including friction stir welding), magnetic pulse welding, co-extrusion welding, cold welding, diffusion bonding, exothermic welding, high frequency welding, hot pressure welding, induction welding, and roll welding.

Geometry

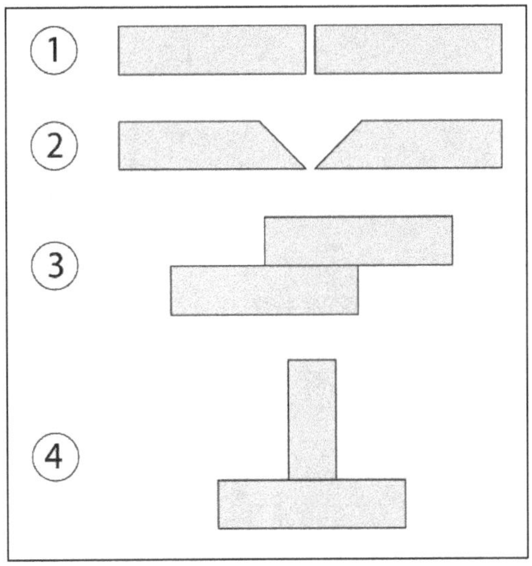

Common welding joint types – (1) Square butt joint, (2) V butt joint, (3) Lap joint, (4) T-joint.

Welds can be geometrically prepared in many different ways. The five basic types of weld joints are the butt joint, lap joint, corner joint, edge joint, and T-joint (a variant of this last is the cruciform joint). Other variations exist as well—for example, double-V preparation joints are characterized by the two pieces of material each tapering to a single center point at one-half their height. Single-U and double-U preparation joints are also fairly common—instead of having straight edges like the single-V and double-V preparation joints, they are curved, forming the shape of a U. Lap joints are also commonly more than two pieces thick—depending on the process used and the thickness of the material, many pieces can be welded together in a lap joint geometry.

Many welding processes require the use of a particular joint design; for example, resistance spot welding, laser beam welding, and electron beam welding are most frequently performed on lap joints. Other welding methods, like shielded metal arc welding, are extremely versatile and can weld virtually any type of joint. Some processes can also be used to make multipass welds, in which one weld is allowed to cool, and then another weld is performed on top of it. This allows for the welding of thick sections arranged in a single-V preparation joint, for example.

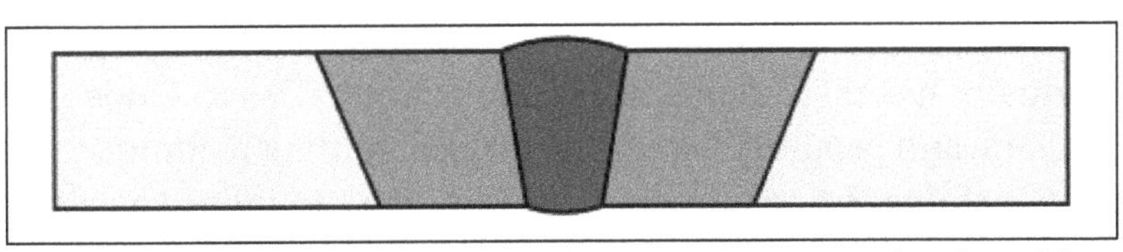

The cross-section of a welded butt joint, with the darkest gray representing the weld or fusion zone, the medium gray the heat-affected zone, and the lightest gray the base material.

After welding, a number of distinct regions can be identified in the weld area. The weld itself is called the fusion zone—more specifically, it is where the filler metal was laid during the welding process. The properties of the fusion zone depend primarily on the filler metal used, and its compatibility with the base materials. It is surrounded by the heat-affected zone, the area that had its microstructure and properties altered by the weld. These properties depend on the base material's behavior when subjected to heat. The metal in this area is often weaker than both the base material and the fusion zone, and is also where residual stresses are found.

Quality

The blue area results from oxidation at a corresponding temperature of 600 °F (316 °C). This is an accurate way to identify temperature, but does not represent the HAZ width. The HAZ is the narrow area that immediately surrounds the welded base metal.

Many distinct factors influence the strength of welds and the material around them, including the welding method, the amount and concentration of energy input, the weldability of the base material, filler material, and flux material, the design of the joint, and the interactions between all these factors. To test the quality of a weld, either destructive or nondestructive testing methods are commonly used to verify that welds are free of defects, have acceptable levels of residual stresses and distortion, and have acceptable heat-affected zone (HAZ) properties. Types of welding defects include cracks, distortion, gas inclusions (porosity), non-metallic inclusions, lack of fusion, incomplete penetration, lamellar tearing, and undercutting.

The metalworking industry has instituted specifications and codes to guide welders, weld inspectors, engineers, managers, and property owners in proper welding technique, design of welds, how to judge the quality of Welding Procedure Specification, how to judge the skill of the person performing the weld, and how to ensure the quality of a welding job. Methods such as visual inspection, radiography, ultrasonic testing, phased-array ultrasonics, dye penetrant inspection, magnetic particle inspection, or industrial computed tomography can help with detection and analysis of certain defects.

Heat-affected Zone

The effects of welding on the material surrounding the weld can be detrimental—depending on the materials used and the heat input of the welding process used, the HAZ can be of varying size and strength. The thermal diffusivity of the base material plays a large role—if the diffusivity is high, the material cooling rate is high and the HAZ is relatively small. Conversely, a low diffusivity leads to slower cooling and a larger HAZ. The amount of heat injected by the welding process plays an important role as well, as processes like oxyacetylene welding have an unconcentrated heat input and increase the size of the HAZ. Processes like laser beam welding give a highly concentrated, limited amount of heat, resulting in a small HAZ. Arc welding falls between these two extremes, with the individual processes varying somewhat in heat input. To calculate the heat input for arc welding procedures, the following formula can be used:

$$Q = \left(\frac{V \times I \times 60}{S \times 1000} \right) \times \mathit{Efficiency}$$

where Q = heat input (kJ/mm), V = voltage (V), I = current (A), and S = welding speed (mm/min). The efficiency is dependent on the welding process used, with shielded metal arc welding having a value of 0.75, gas metal arc welding and submerged arc welding, 0.9, and gas tungsten arc welding, 0.8.

Lifetime Extension with Aftertreatment Methods

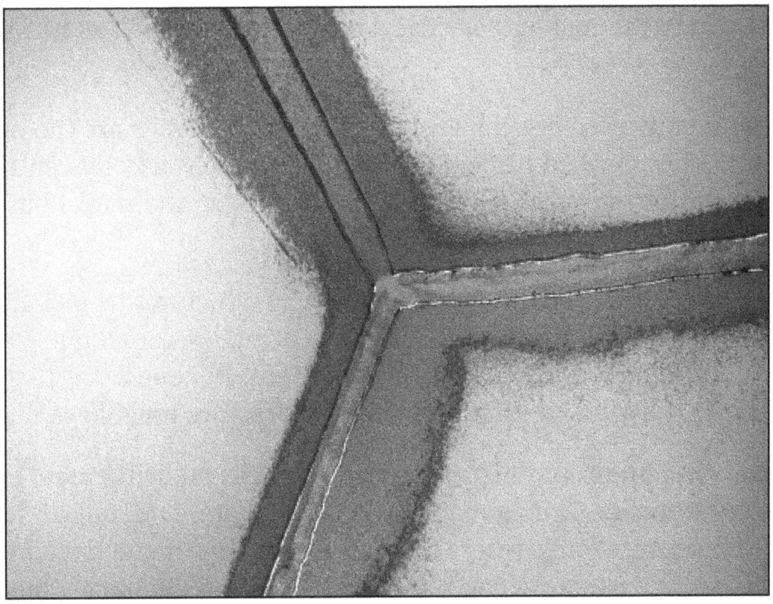

Example: High Frequency Impact Treatment for lifetime extension.

The durability and life of dynamically loaded, welded steel structures is determined in many cases by the welds, particular the weld transitions. Through selective treatment of the transitions by grinding (abrasive cutting), shot peening, High Frequency Impact Treatment, etc. the durability of many designs increase significantly.

Metallurgy

Most solids used are engineering materials consisting of crystalline solids in which the atoms or ions are arranged in a repetitive geometric pattern which is known as a lattice structure. The only exception is material that is made from glass which is a combination of a supercooled liquid and polymers which are aggregates of large organic molecules.

Crystalline solids cohesion is obtained by a metallic or chemical bond which is formed between the constituent atoms. Chemical bonds can be grouped into two types consisting of ionic and co-valent. To form an ionic bond, either a valence or bonding electron separates from one atom and becomes attached to another atom to form oppositely charged ions. The bonding in the static position is when the ions occupy an equilibrium position where the resulting force between them is zero. When the ions are exerted in tension force, the inter-ionic spacing increases creating an electrostatic attractive force, while a repulsing force under compressive force between the atomic nuclei is dominant.

Covalent bonding takes place when one of the constituent atoms loses one or more electrons, with the other atom gaining the electrons, resulting in an electron cloud that is shared by the molecule as a whole. In both ionic and covalent bonding the location of the ions and electrons are constrained relative to each other, thereby resulting in the bond being characteristically brittle.

Metallic bonding can be classified as a type of covalent bonding for which the constituent atoms of the same type and do not combine with one another to form a chemical bond. Atoms will lose an electron(s) forming an array of positive ions. These electrons are shared by the lattice which makes the electron cluster mobile, as the electrons are free to move as well as the ions. For this, it gives metals their relatively high thermal and electrical conductivity as well as being characteristically ductile.

Three of the most commonly used crystal lattice structures in metals are the body-centred cubic, face-centred cubic and close-packed hexagonal. Ferritic steel has a body-centred cubic structure and austenitic steel, non-ferrous metals like aluminum, copper and nickel have the face-centred cubic structure.

Ductility is an important factor in ensuring the integrity of structures by enabling them to sustain local stress concentrations without fracture. In addition, structures are required to be of an acceptable strength, which is related to a material's yield strength. In general, as the yield strength of a material increases, there is a corresponding reduction in fracture toughness.

A reduction in fracture toughness may also be attributed to the embrittlement effect of impurities, or for body-centred cubic metals, from a reduction in temperature. Metals and in particular steels have a transitional temperature range where above this range the metal has acceptable notch-ductility while below this range the material becomes brittle. Within the range, the materials behavior is unpredictable. The reduction in fracture toughness is accompanied by a change in the fracture appearance. When above the transition, the fracture is primarily due to micro-void coalescence, which results in the fracture appearing fibrous. When the temperatures falls the fracture will show signs of cleavage facets. These two appearances are visible by the naked eye. Brittle fracture in steel plates may appear as chevron markings under the microscope. These arrow-like ridges on the crack surface point towards the origin of the fracture.

Fracture toughness is measured using a notched and pre-cracked rectangular specimen, of which the dimensions are specified in standards, for example ASTM E23. There are other means of estimating or measuring fracture toughness by the following: The Charpy impact test per ASTM A370; The crack-tip opening displacement (CTOD) test per BS 7448-1; The J integral test per ASTM E1820; The Pellini drop-weight test per ASTM E208.

Unusual Conditions

Underwater welding.

While many welding applications are done in controlled environments such as factories and repair shops, some welding processes are commonly used in a wide variety of conditions, such as open air, underwater, and vacuums (such as space). In open-air applications, such as construction and outdoors repair, shielded metal arc welding is the most common process. Processes that employ inert gases to protect the weld cannot be readily used in such situations, because unpredictable atmospheric movements can result in a faulty weld. Shielded metal arc welding is also often used in underwater welding in the construction and repair of ships, offshore platforms, and pipelines, but others, such as flux cored arc welding and gas tungsten arc welding, are also common. Welding in space is also possible—it was first attempted in 1969 by Russian cosmonauts, when they performed experiments to test shielded metal arc welding, plasma arc welding, and electron beam welding in a depressurized environment. Further testing of these methods was done in the following decades, and today researchers continue to develop methods for using other welding processes in space, such as laser beam welding, resistance welding, and friction welding. Advances in these areas may be useful for future endeavours similar to the construction of the International Space Station, which could rely on welding for joining in space the parts that were manufactured on Earth.

Safety Issues

Welding can be dangerous and unhealthy if the proper precautions are not taken. However, using new technology and proper protection greatly reduces risks of injury and death associated with welding. Since many common welding procedures involve an open electric arc or flame, the risk of burns and fire is significant; this is why it is classified as a hot work process. To prevent injury, welders wear personal protective equipment in the form of heavy leather gloves and protective long-sleeve jackets to avoid exposure to extreme heat and flames. Additionally, the brightness of the weld area leads to a condition called arc eye or flash burns in which ultraviolet light causes inflammation of the cornea and can burn the retinas of the eyes. Goggles and welding helmets with dark UV-filtering face plates are worn to prevent this exposure. Since the 2000s, some helmets have included a face plate which instantly darkens upon exposure to the intense UV light. To protect bystanders, the welding area is often surrounded with translucent welding curtains. These curtains, made of a polyvinyl chloride plastic film, shield people outside the welding area from the UV light of the electric arc, but can not replace the filter glass used in helmets.

Arc welding with a welding helmet, gloves, and other protective clothing.

A chamber designed to contain welding fumes for analysis.

Welders are often exposed to dangerous gases and particulate matter. Processes like flux-cored arc welding and shielded metal arc welding produce smoke containing particles of various types of oxides. The size of the particles in question tends to influence the toxicity of the fumes, with smaller particles presenting a greater danger. This is because smaller particles have the ability to cross the blood brain barrier. Fumes and gases, such as carbon dioxide, ozone, and fumes containing heavy metals, can be dangerous to welders lacking proper ventilation and training. Exposure to manganese welding fumes, for example, even at low levels ($<0.2\ mg/m^3$), may lead to neurological problems or to damage to the lungs, liver, kidneys, or central nervous system. Nano particles can become trapped in the alveolar macrophages of the lungs and induce pulmonary fibrosis. The use of compressed gases and flames in many welding processes poses an explosion and fire risk. Some common precautions include limiting the amount of oxygen in the air, and keeping combustible materials away from the workplace.

Costs and Trends

As an industrial process, the cost of welding plays a crucial role in manufacturing decisions. Many different variables affect the total cost, including equipment cost, labor cost, material cost, and energy cost. Depending on the process, equipment cost can vary, from inexpensive for methods like shielded metal arc welding and oxyfuel welding, to extremely expensive for methods like laser beam welding and electron beam welding. Because of their high cost, they are only used in high production operations. Similarly, because automation and robots increase equipment costs, they are only implemented when high production is necessary. Labor cost depends on the deposition rate (the rate of welding), the hourly wage, and the total operation time, including time spent fitting, welding, and handling the part. The cost of materials includes the cost of the base and filler material, and the cost of shielding gases. Finally, energy cost depends on arc time and welding power demand.

For manual welding methods, labor costs generally make up the vast majority of the total cost. As a result, many cost-saving measures are focused on minimizing operation time. To do this, welding procedures with high deposition rates can be selected, and weld parameters can be fine-tuned to increase welding speed. Mechanization and automation are often implemented to reduce labor costs, but this frequently increases the cost of equipment and creates additional setup time. Material costs tend to increase when special properties are necessary, and energy costs normally do not amount to more than several percent of the total welding cost.

In recent years, in order to minimize labor costs in high production manufacturing, industrial welding has become increasingly more automated, most notably with the use of robots in resistance spot welding (especially in the automotive industry) and in arc welding. In robot welding, mechanized devices both hold the material and perform the weld and at first, spot welding was its most common application, but robotic arc welding increases in popularity as technology advances. Other key areas of research and development include the welding of dissimilar materials (such as steel and aluminum, for example) and new welding processes, such as friction stir, magnetic pulse, conductive heat seam, and laser-hybrid welding. Furthermore, progress is desired in making more specialized methods like laser beam welding practical for more applications, such as in the aerospace and automotive industries. Researchers also hope to better understand the often unpredictable properties of welds, especially microstructure, residual stresses, and a weld's tendency to crack or deform.

The trend of accelerating the speed at which welds are performed in the steel erection industry comes at a risk to the integrity of the connection. Without proper fusion to the base materials provided by sufficient arc time on the weld, a project inspector cannot ensure the effective diameter of the puddle weld therefore he or she cannot guarantee the published load capacities unless they witness the actual installation. This method of puddle welding is common in the United States and Canada for attaching steel sheets to bar joist and structural steel members. Regional agencies are responsible for ensuring the proper installation of puddle welding on steel construction sites. Currently there is no standard or weld procedure which can ensure the published holding capacity of any unwitnessed connection, but this is under review by the American Welding Society.

Glass and Plastic Welding

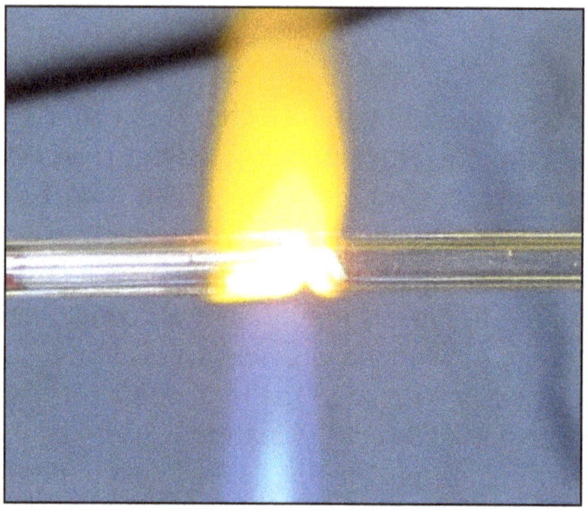

The welding together of two tubes made from lead glass.

Glasses and certain types of plastics are commonly welded materials. Unlike metals, which have a specific melting point, glasses and plastics have a melting range, called the glass transition. When heating the solid material into this range, it will generally become softer and more pliable. When it crosses through the glass transition, it will become a very thick, sluggish, viscous liquid. Typically, this viscous liquid will have very little surface tension, becoming a sticky, honey-like consistency, so welding can usually take place by simply pressing two melted surfaces together. The two liquids will generally mix and join at first contact. Upon cooling through the glass transition, the welded piece will solidify as one solid piece of amorphous material.

A bowl made from cast-glass. The two halves are joined together by the weld seam, running down the middle.

Glass Welding

Glass welding is a common practice during glassblowing. It is used very often in the construction of lighting, neon signs, flashtubes, scientific equipment, and the manufacture of dishes and other glassware. It is also used during glass casting for joining the halves of glass molds, making items such as bottles and jars. Welding glass is accomplished by heating the glass through the glass transition, turning it into a thick, formable, liquid mass. Heating is usually done with a gas or oxy-gas torch, or a furnace, because the temperatures for melting glass are often quite high. This temperature may vary, depending on the type of glass. For example, lead glass becomes a weldable liquid at around 1,600 °F (870 °C), and can be welded with a simple propane torch. On the other hand, quartz glass (fused silica) must be heated to over 3,000 °F (1,650 °C), but quickly loses its viscosity and formability if overheated, so an oxyhydrogen torch must be used. Sometimes a tube may be attached to the glass, allowing it to be blown into various shapes, such as bulbs, bottles, or tubes. When two pieces of liquid glass are pressed together, they will usually weld very readily. Welding a handle onto a pitcher can usually be done with relative ease. However, when welding a tube to another tube, a combination of blowing and suction, and pressing and pulling is used to ensure a good seal, to shape the glass, and to keep the surface tension from closing the tube in on itself. Sometimes a filler rod may be used, but usually not.

Because glass is very brittle in its solid state, it is often prone to cracking upon heating and cooling, especially if the heating and cooling are uneven. This is because the brittleness of glass does not allow for uneven thermal expansion. Glass that has been welded will usually need to be cooled very slowly and evenly through the glass transition, in a process called annealing, to relieve any internal stresses created by a temperature gradient.

There are many types of glass, and it is most common to weld using the same types. Different glasses often have different rates of thermal expansion, which can cause them to crack upon cooling when they contract differently. For instance, quartz has very low thermal expansion, while soda-lime glass has very high thermal expansion. When welding different glasses to each other, it is usually important to closely match their coefficients of thermal expansion, to ensure that cracking does not occur. Also, some glasses will simply not mix with others, so welding between certain types may not be possible.

Glass can also be welded to metals and ceramics, although with metals the process is usually more adhesion to the surface of the metal rather than a commingling of the two materials. However, certain glasses will typically bond only to certain metals. For example, lead glass bonds readily to copper or molybdenum, but not to aluminum. Tungsten electrodes are often used in lighting but will not bond to quartz glass, so the tungsten is often wetted with molten borosilicate glass, which bonds to both tungsten and quartz. However, care must be taken to ensure that all materials have similar coefficients of thermal expansion to prevent cracking both when the object cools and when it is heated again. Special alloys are often used for this purpose, ensuring that the coefficients of expansion match, and sometimes thin, metallic coatings may be applied to a metal to create a good bond with the glass.

Plastic Welding

Plastics are generally divided into two categories, which are "thermosets" and "thermoplastics." A thermoset is a plastic in which a chemical reaction sets the molecular bonds after first forming the plastic, and then the bonds cannot be broken again without degrading the plastic. Thermosets cannot be melted, therefore, once a thermoset has set it is impossible to weld it. Examples of thermosets include epoxies, silicone, vulcanized rubber, polyester, and polyurethane.

Thermoplastics, by contrast, form long molecular chains, which are often coiled or intertwined, forming an amorphous structure without any long-range, crystalline order. Some thermoplastics may be fully amorphous, while others have a partially crystalline/partially amorphous structure. Both amorphous and semicrystalline thermoplastics have a glass transition, above which welding can occur, but semicrystallines also have a specific melting point which is above the glass transition. Above this melting point, the viscous liquid will become a free-flowing liquid. Examples of thermoplastics include polyethylene, polypropylene, polystyrene, polyvinylchloride (PVC), and fluoroplastics like Teflon and Spectralon.

Welding thermoplastic is very similar to welding glass. The plastic first must be cleaned and then heated through the glass transition, turning the weld-interface into a thick, viscous liquid. Two heated interfaces can then be pressed together, allowing the molecules to mix through intermolecular diffusion, joining them as one. Then the plastic is cooled through the glass transition, allowing the weld to solidify. A filler rod may often be used for certain types of joints. The main differences between welding glass and plastic are the types of heating methods, the much lower melting temperatures, and the fact that plastics will burn if overheated. Many different methods have been devised for heating plastic to a weldable temperature without burning it. Ovens or electric heating tools can be used to melt the plastic. Ultrasonic, laser, or friction heating are other methods. Resistive metals may be implanted in the plastic, which respond to induction heating. Some plastics will begin to burn at temperatures lower than their glass transition, so welding can be performed by blowing a heated, inert gas onto the plastic, melting it while, at the same time, shielding it from oxygen.

Many thermoplastics can also be welded using chemical solvents. When placed in contact with the plastic, the solvent will begin to soften it, bringing the surface into a thick, liquid solution. When two melted surfaces are pressed together, the molecules in the solution mix, joining them as one. Because the solvent can permeate the plastic, the solvent evaporates out through the surface of the plastic, causing the weld to drop out of solution and solidify. A common use for solvent welding is for joining PVC or ABS (acrylonitrile butadiene styrene) pipes during plumbing, or for welding styrene and polystyrene plastics in the construction of models. Solvent welding is especially effective on plastics like PVC which burn at or below their glass transition, but may be ineffective on plastics like Teflon or polyethylene that are resistant to chemical decomposition.

Annealing

Annealing, in metallurgy and materials science, is a heat treatment that alters the physical and sometimes chemical properties of a material to increase its ductility and reduce its hardness, making it more workable. It involves heating a material to above its recrystallization temperature, maintaining a suitable temperature, and then cooling.

In annealing, atoms migrate in the crystal lattice and the number of dislocations decreases, leading to the change in ductility and hardness.

In the cases of copper, steel, silver, and brass, this process is performed by heating the material (generally until glowing) for a while and then slowly letting it cool to room temperature in still air. Copper, silver and brass can be cooled slowly in air, or quickly by quenching in water, unlike ferrous metals, such as steel, which must be cooled slowly to anneal. In this fashion, the metal is softened and prepared for further work—such as shaping, stamping, or forming.

Thermodynamics

Annealing occurs by the diffusion of atoms within a solid material, so that the material progresses towards its equilibrium state. Heat increases the rate of diffusion by providing the energy needed to break bonds. The movement of atoms has the effect of redistributing and eradicating the dislocations in metals and (to a lesser extent) in ceramics. This alteration to existing dislocations allows a metal object to deform more easily, increasing its ductility.

The amount of process-initiating Gibbs free energy in a deformed metal is also reduced by the annealing process. In practice and industry, this reduction of Gibbs free energy is termed *stress relief.*

The relief of internal stresses is a thermodynamically spontaneous process; however, at room temperatures, it is a very slow process. The high temperatures at which annealing occurs serve to accelerate this process.

The reaction that facilitates returning the cold-worked metal to its stress-free state has many reaction pathways, mostly involving the elimination of lattice vacancy gradients within the body of the metal. The creation of lattice vacancies is governed by the Arrhenius equation, and the migration/ diffusion of lattice vacancies are governed by Fick's laws of diffusion.

In steel, there is a decarburation mechanism that can be described as three distinct events: the reaction at the steel surface, the interstitial diffusion of carbon atoms and the dissolution of carbides within the steel.

Stages

The three stages of the annealing process that proceed as the temperature of the material is increased are: recovery, recrystallization, and grain growth. The first stage is recovery, and it results in softening of the metal through removal of primarily linear defects called *dislocations* and the internal stresses they cause. Recovery occurs at the lower temperature stage of all annealing processes and before the appearance of new strain-free grains. The grain size and shape do not change. The second stage is recrystallization, where new strain-free grains nucleate and grow to replace those deformed by internal stresses. If annealing is allowed to continue once recrystallization has completed, then grain growth (the third stage) occurs. In grain growth, the microstructure starts to coarsen and may cause the metal to lose a substantial part of its original strength. This can however be regained with hardening.

Controlled Atmospheres

The high temperature of annealing may result in oxidation of the metal's surface, resulting in scale.

If scale must be avoided, annealing is carried out in a special atmosphere, such as with endothermic gas (a mixture of carbon monoxide, hydrogen gas, and nitrogen gas). Annealing is also done in forming gas, a mixture of hydrogen and nitrogen.

The magnetic properties of mu-metal (Espey cores) are introduced by annealing the alloy in a hydrogen atmosphere.

Setup and Equipment

Typically, large ovens are used for the annealing process. The inside of the oven is large enough to place the workpiece in a position to receive maximum exposure to the circulating heated air. For high volume process annealing, gas fired conveyor furnaces are often used. For large workpieces or high quantity parts, car-bottom furnaces are used so workers can easily move the parts in and out. Once the annealing process is successfully completed, workpieces are sometimes left in the oven so the parts cool in a controllable way. While some workpieces are left in the oven to cool in a controlled fashion, other materials and alloys are removed from the oven. Once removed from the oven, the workpieces are often quickly cooled off in a process known as quench hardening. Typical methods of quench hardening materials involve media such as air, water, oil, or salt. Salt is used as a medium for quenching usually in the form of brine (salt water). Brine provides faster cooling rates than water. This is because when an object is quenched in water air bubbles form on the surface of the object reducing the surface area the water is in contact with. The salt in the brine reduces the formation of air bubbles on the object's surface, meaning there is a larger surface area of the object in contact with the water, providing faster cooling rates. Quench hardening is generally applicable to some ferrous alloys, but not copper alloys.

Diffusion Annealing of Semiconductors

In the semiconductor industry, silicon wafers are annealed, so that dopant atoms, usually boron, phosphorus or arsenic, can diffuse into substitutional positions in the crystal lattice, resulting in drastic changes in the electrical properties of the semiconducting material.

Specialized Cycles

Normalization

Normalization is an annealing process applied to ferrous alloys to give the material a uniform fine-grained structure and make it less brittle. It is used on steels of less than 0.4% carbon to transform austenite into ferrite, pearlite and sorbite. It involves heating the steel to 20-50 Kelvin above its upper critical point. It is soaked for a short period at that temperature and then allowed to cool in air. Smaller grains form that produce a tougher, more ductile material. It eliminates columnar grains and dendritic segregation that sometimes occurs during casting. Normalizing improves machinability of a component and provides dimensional stability if subjected to further heat treatment processes.

Process Annealing

Process annealing, also called *intermediate annealing*, *subcritical annealing*, or *in-process annealing*, is a heat treatment cycle that restores some of the ductility to a product being cold-worked so it can be cold-worked further without breaking.

The temperature range for process annealing ranges from 260 °C (500 °F) to 760 °C (1400 °F), depending on the alloy in question. This process is mainly suited for low-carbon steel. The material is heated up to a temperature just below the lower critical temperature of steel. Cold-worked steel normally tends to possess increased hardness and decreased ductility, making it difficult to work. Process annealing tends to improve these characteristics. This is mainly carried out on cold-rolled steel like wire-drawn steel, etc.

Full Anneal

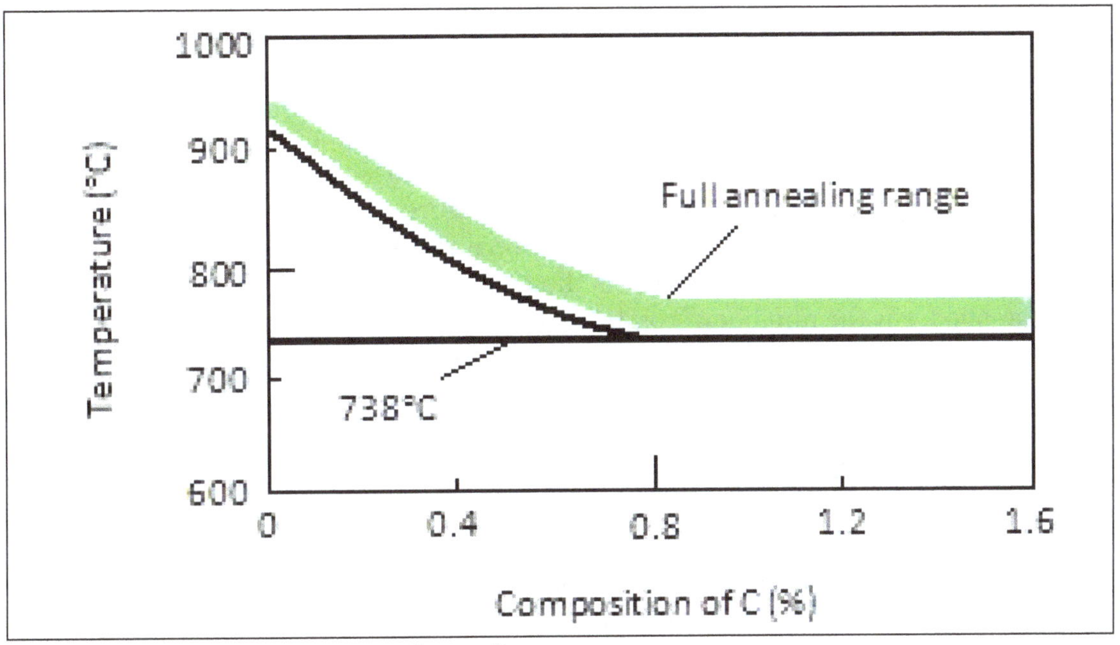

Full annealing temperature ranges.

A full anneal typically results in the second most ductile state a metal can assume for metal alloy. Its purpose is to originate a uniform and stable microstructure that most closely resembles the metal's phase diagram equilibrium microstructure, thus letting the metal attain relatively low levels of hardness, yield strength and ultimate strength with high plasticity and toughness. To perform a full anneal on a steel for example, steel is heated to slightly above the austenitic temperature and held for sufficient time to allow the material to fully form austenite or austenite-cementite grain structure. The material is then allowed to cool very slowly so that the equilibrium microstructure is obtained. In most cases this means the material is allowed to furnace cool (the furnace is turned off and the steel is let cool down inside) but in some cases it's air cooled. The cooling rate of the steel has to be sufficiently slow so as to not let the austenite transform into bainite or martensite, but rather have it completely transform to pearlite and ferrite or cementite. This means that steels that are very hardenable (i.e. tend to form martensite under moderately low cooling rates) have to be furnace cooled. The details of the process depend on the type of metal and the precise alloy involved. In any case the result is a more ductile material but a lower yield strength and a lower tensile strength. This process is also called LP annealing for *lamellar pearlite* in the steel industry as opposed to a *process anneal*, which does not specify a microstructure and only has the goal of softening the material. Often the material to be machined is annealed, and then subject to further heat treatment to achieve the final desired properties.

Short Cycle Anneal

Short cycle annealing is used for turning normal ferrite into malleable ferrite. It consists of heating, cooling and then heating again from 4 to 8 hours.

Resistive Heating

Resistive heating can be used to efficiently anneal copper wire; the heating system employs a controlled electrical short circuit. It can be advantageous because it does not require a temperature-regulated furnace like other methods of annealing.

The process consists of two conductive pulleys (*step pulleys*), which the wire passes across after it is drawn. The two pulleys have an electrical potential across them, which causes the wire to form a short circuit. The Joule effect causes the temperature of the wire to rise to approximately 400 °C. This temperature is affected by the rotational speed of the pulleys, the ambient temperature, and the voltage applied. Where t is the temperature of the wire, K is a constant, V is the voltage applied, r is the number of rotations of the pulleys per minute, and t_a is the ambient temperature:

$$t = \frac{1}{r} KV^2 + t_a$$

The constant K depends on the diameter of the pulleys and the resistivity of the copper.

Purely in terms of the temperature of the copper wire, an increase in the speed of the wire through the pulley system has the same effect as an increase in resistance. Therefore, the speed of the wire can be varied quadratically as the voltage is applied.

Calcination

Authorities differ on the meaning of calcination (also referred to as calcining). The IUPAC defines it as 'Heating to high temperatures in air or oxygen'. However calcination is also used to mean a thermal treatment process in the absence or limited supply of air or oxygen applied to ores and other solid materials to bring about a thermal decomposition A calciner is a steel cylinder that rotates inside a heated furnace and performs indirect high-temperature processing (550-1150 °C, or 1000-2100 °F) within a controlled atmosphere.

Industrial Processes

The process of calcination derives its name from the Latin *calcinare* (to burn lime) due to its most common application, the decomposition of calcium carbonate (limestone) to calcium oxide (lime) and carbon dioxide, in order to create cement. The product of calcination is usually referred to in general as "calcine", regardless of the actual minerals undergoing thermal treatment. Calcination is carried out in furnaces or reactors (sometimes referred to as kilns or calciners) of various designs including shaft furnaces, rotary kilns, multiple hearth furnaces, and fluidized bed reactors.

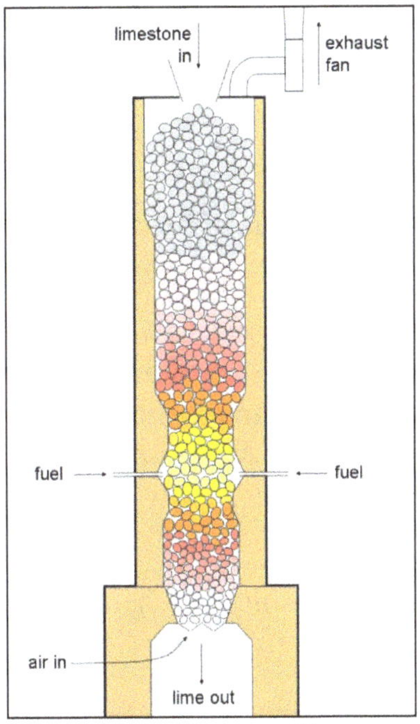

An oven for calcination of limestone.

Examples of calcination processes include the following:

- decomposition of carbonate minerals, as in the calcination of limestone to drive off carbon dioxide;
- decomposition of hydrated minerals, as in the calcination of bauxite and gypsum, to remove crystalline water as water vapor;
- decomposition of volatile matter contained in raw petroleum coke;
- heat treatment to effect phase transformations, as in conversion of anatase to rutile or devitrification of glass materials;
- removal of ammonium ions in the synthesis of zeolites.

Calcination Reactions

Calcination reactions usually take place at or above the thermal decomposition temperature (for decomposition and volatilization reactions) or the transition temperature (for phase transitions). This temperature is usually defined as the temperature at which the standard Gibbs free energy for a particular calcination reaction is equal to zero. For example, in limestone calcination, a decomposition process, the chemical reaction is,

$$CaCO_3 \rightarrow CaO + CO_2(g)$$

The standard Gibbs free energy of reaction is approximated as $\Delta G°_r = 177{,}100 - 158\ T$ (J/mol). The standard free energy of reaction is 0 in this case when the temperature, T, is equal to 1121 K, or 848 °C.

Oxidation

In some cases, calcination of a metal results in oxidation of the metal. Jean Rey noted that lead and tin when calcinated gained weight, presumably as they were being oxidized.

Alchemy

In alchemy, calcination was believed to be one of the 12 vital processes required for the transformation of a substance.

Alchemists distinguished two kinds of calcination, *actual* and *potential*. Actual calcination is that brought about by actual fire, from wood, coals, or other fuel, raised to a certain temperature. Potential calcination is that brought about by *potential* fire, such as corrosive chemicals; for example, gold was calcined in a reverberatory furnace with mercury and sal ammoniac; silver with common salt and alkali salt; copper with salt and sulfur; iron with sal ammoniac and vinegar; tin with antimony; lead with sulfur; and mercury with aqua fortis.

There was also *philosophical calcination*, which was said to occur when horns, hooves, etc., were hung over boiling water, or other liquor, until they had lost their mucilage, and were easily reducible into powder.

Recovery

Recovery is a process by which deformed grains can reduce their stored energy by the removal or rearrangement of defects in their crystal structure. These defects, primarily dislocations, are introduced by plastic deformation of the material and act to increase the yield strength of a material. Since recovery reduces the dislocation density the process is normally accompanied by a reduction in a materials strength and a simultaneous increase in the ductility. As a result, recovery may be considered beneficial or detrimental depending on the circumstances. Recovery is related to the similar process of recrystallization and grain growth. Recovery competes with recrystallization, as both are driven by the stored energy, but is also thought to be a necessary prerequisite for the nucleation of recrystallized grains. It is so called because there is a recovery of the electrical conductivity due to a reduction in dislocations. This creates defect-free channels, giving electrons an increased mean-free path.

Definition

The physical processes that fall under the designations of recovery, recrystallisation and grain growth are often difficult to distinguish in a precise manner. Doherty *et al.* (1998) stated:

"The authors have agreed that ... recovery can be defined as all annealing processes occurring in deformed materials that occur without the migration of a high-angle grain boundary"

Thus the process can be differentiated from recrystallisation and grain growth as both feature extensive movement of high-angle grain boundaries.

If recovery occurs during deformation (a situation that is common in high-temperature processing) then it is referred to as 'dynamic' while recovery that occurs after processing is termed 'static'. The principal difference is that during dynamic recovery, stored energy continues to be introduced even as it is decreased by the recovery process - resulting in a form of dynamic equilibrium.

Process

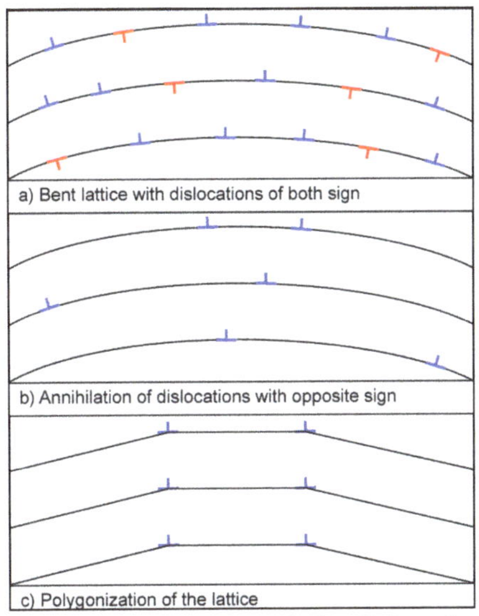

The annihilation and reorganisation of an array of edge dislocations in a crystal lattice.

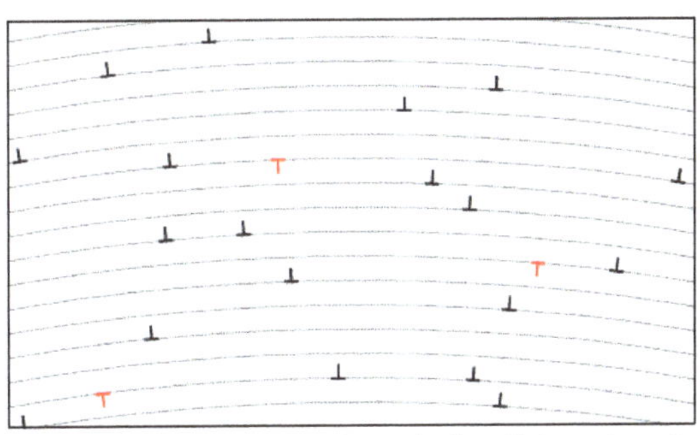

Animation of the annihilation and reorganisation of edge dislocations in a crystal lattice.

Deformed Structure

A heavily deformed metal contains a huge number of dislocations predominantly caught up in 'tangles' or 'forests'. Dislocation motion is relatively difficult in a metal with a low stacking fault energy and so the dislocation distribution after deformation is largely random. In contrast, metals with moderate to high stacking fault energy, e.g. aluminum, tend to form a cellular structure where the cell walls consist of rough tangles of dislocations. The interiors of the cells have a correspondingly reduced dislocation density.

Annihilation

Each dislocation is associated with a strain field which contributes some small but finite amount to the materials stored energy. When the temperature is increased - typically below one-third of the absolute melting point - dislocations become mobile and are able to glide, cross-slip and climb. If two dislocations of opposite sign meet then they effectively cancel out and their contribution to the stored energy is removed. When annihilation is complete then only excess dislocation of one kind will remain.

Rearrangement

After annihilation any remaining dislocations can align themselves into ordered arrays where their individual contribution to the stored energy is reduced by the overlapping of their strain fields. The simplest case is that of an array of edge dislocations of identical Burger's vector. This idealised case can be produced by bending a single crystal that will deform on a single slip system (the original experiment performed by Cahn in 1949). The edge dislocations will rearrange themselves into tilt boundaries, a simple example of a low-angle grain boundary. Grain boundary theory predicts that an increase in boundary misorientation will increase the energy of the boundary but decrease the energy per dislocation. Thus, there is a driving force to produce fewer, more highly misoriented boundaries. The situation in highly deformed, polycrystalline materials is naturally more complex. Many dislocations of different Burger's vector can interact to form complex 2-D networks.

Development of Substructure

As mentioned above, the deformed structure is often a 3-D cellular structure with walls consisting of dislocation tangles. As recovery proceeds these cell walls will undergo a transition towards a genuine subgrain structure. This occurs through a gradual elimination of extraneous dislocations and the rearrangement of the remaining dislocations into low-angle grain boundaries.

Sub-grain formation is followed by subgrain coarsening where the average size increases while the number of subgrains decreases. This reduces the total area of grain boundary and hence the stored energy in the material. Subgrain coarsen shares many features with grain growth.

If the sub-structure can be approximated to an array of spherical subgrains of radius R and boundary energy γ_s; the stored energy is uniform; and the force on the boundary is evenly distributed, the driving pressure P is given by:

$$P = -\alpha R \frac{d}{dR}\left(\frac{\gamma_s}{R}\right)$$

Since γ_s is dependent on the boundary misorientation of the surrounding subgrains, the driving pressure generally does not remain constant throughout coarsening.

Roasting

Roasting is a step of the processing of certain ores. More specifically, roasting is a metallurgical process involving gas–solid reactions at elevated temperatures with the goal of purifying the metal component(s). Often before roasting, the ore has already been partially purified, e.g. by froth flotation. The concentrate is mixed with other materials to facilitate the process. The technology is useful but is also a serious source of air pollution.

Roasted gold ore from Cripple Creek, Colorado. Roasting has driven off the tellurium from the original calaverite, leaving behind vesicular blebs of native gold.

Roasting consists of thermal gas–solid reactions, which can include oxidation, reduction, chlorination, sulfation, and pyrohydrolysis. In roasting, the ore or ore concentrate is treated with very hot air. This process is generally applied to sulfide minerals. During roasting, the sulfide is converted to an oxide, and sulfur is released as sulfur dioxide, a gas. For the ores Cu_2S (chalcocite) and ZnS (sphalerite), balanced equations for the roasting are:

$$2\ Cu_2S + 3O_2 \rightarrow 2\ Cu_2O + 2\ SO_2$$

$$2\ ZnS + 3\ O_2 \rightarrow 2\ ZnO + 2\ SO_2$$

The gaseous product of sulfide roasting, sulfur dioxide (SO_2) is often used to produce sulfuric acid. Many sulfide minerals contain other components such as arsenic that are released into the environment.

Up until the early 20th century, roasting was started by burning wood on top of ore. This would raise the temperature of the ore to the point where its sulfur content would become its source of fuel, and the roasting process could continue without external fuel sources. Early sulfide roasting was practiced in this manner in "open hearth" roasters, which were manually stirred (a practice referred to as "rabbling") using rake-like tools to expose unroasted ore to oxygen as the reaction proceeded.

This process would release large amounts of acidic, metallic, and other toxic compounds. Results of this include areas that even after 60-80 years are still largely lifeless, often exactly corresponding to the area of the roast bed, some of which are hundreds of metres wide by kilometres long.

Roasting Operations

Oxidizing Roasting

Oxidizing roasting, the most commonly practiced roasting process, involves heating the ore in excess of air or oxygen, to burn out or replace the impurity element, generally sulfur, partly or completely by oxygen. For sulfide roasting, the general reaction can be given by -

$$MS\ (s) + 1.5\ O_2\ (g) = MO\ (s) + SO_2\ (g)$$

Roasting the sulfide ore, till almost complete removal of the sulfur from the ore, results in a *dead roast*.

Volatilizing Roasting

Volatilizing roasting, involves careful oxidation at elevated temperatures of the ores, to eliminate impurity elements in the form of their volatile oxides. Examples of such volatile oxides include As_2O_3, Sb_2O_3, ZnO and sulfur oxides. Careful control of the oxygen content in the roaster is necessary, as excessive oxidation forms non volatile oxides.

Chloridizing Roasting

Chloridizing roasting transforms certain metal compounds to chlorides, through oxidation or reduction. Some metals like uranium, titanium, beryllium and some rare earths are processed in their chloride form. Certain forms of chloridizing roasting maybe represented by the overall reactions -

$$2NaCl + MS + 2O_2 = Na_2SO_4 + MCl,$$

$$4NaCl + 2MO + S_2 + 3O_2 = 2Na_2SO_4 + 2MCl_2$$

The first reaction represents the chlorination of a sulfide ore involving an exothermic reaction. The second reaction involving an oxide ore is facilitated by addition of elemental sulfur. Carbonate ores react in a similar manner as the oxide ore, after decomposing to their oxide form at high temperature.

Sulfating Roasting

Sulfating roasting oxidizes certain sulfide ores to sulfates in a controlled supply of air to enable leaching of the sulfate for further processing.

Magnetic Roasting

Magnetic roasting involves controlled roasting of the ore to convert it into a magnetic form, thus enabling easy separation and processing in subsequent steps. For example, controlled reduction of haematite (non magnetic Fe_2O_3) to magnetite (magnetic Fe_3O_4).

Reduction Roasting

Reduction roasting partially reduces an oxide ore before the actual smelting process.

Sinter Roasting

Sinter roasting involves heating the fine ores at high temperatures, where simultaneous oxidation and agglomeration of the ores take place. For example, lead sulfide ores are subjected to sinter roasting in a continuous process after froth flotation to convert the fine ores to workable agglomerates for further smelting operations.

Molding

Molding or moulding is the process of manufacturing by shaping liquid or pliable raw material using a rigid frame called a mold or matrix. This itself may have been made using a pattern or model of the final object.

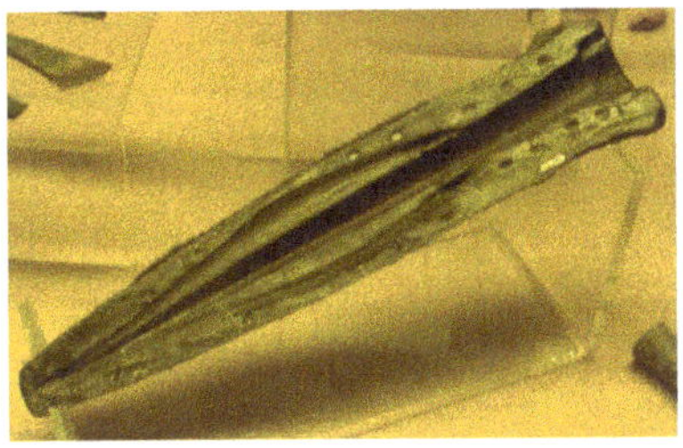

One half of a bronze mold for casting a socketed spear head dated to the period 1400-1000 BC. There are no known parallels for this mold.

Stone mold of the Bronze Age used to produce spear tips.

Ancient Greek molds, used to mass-produce clay figurines, 5th/4th century BC. Beside them, the modern casts taken from them. On display in the Ancient Agora Museum in Athens, housed in the Stoa of Attalus.

Ancient wooden molds used for jaggery & sweets, archaeological museum in Jaffna, Sri Lanka.

A mold or mould is a hollowed-out block that is filled with a liquid or pliable material like plastic, glass, metal, or ceramic raw materials. The liquid hardens or sets inside the mold, adopting its shape.

A mold is the counterpart to a cast. The very common bi-valve molding process uses two molds, one for each half of the object. Piece-molding uses a number of different molds, each creating a section of a complicated object. This is generally only used for larger and more valuable objects.

The manufacturer who makes the molds is called the moldmaker. A release agent is typically used to make removal of the hardened/set substance from the mold easier. Typical uses for molded plastics include molded furniture, molded household goods, molded cases, and structural materials.

Types of molding include:

- Blow molding.
- Powder metallurgy plus sintering.
- Compression molding.
- Extrusion molding.
- Injection molding.
- Laminating.
    - Reaction injection molding.
- Matrix molding.
- Rotational molding (or Rotomolding).
- Spin casting.
- Transfer molding.
- Thermoforming.
    - Vacuum forming, a simplified version of thermoforming.

Electrowinning

Electrorefining technology converting spent commercial nuclear fuel into metal.

Electrowinning, also called electroextraction, is the electrodeposition of metals from their ores that have been put in solution via a process commonly referred to as leaching. Electrorefining uses a similar process to remove impurities from a metal. Both processes use electroplating on a large scale and are important techniques for the economical and straightforward purification of non-ferrous metals. The resulting metals are said to be *electrowon*.

In electrowinning, a current is passed from an inert anode through a liquid *leach* solution containing the metal so that the metal is extracted as it is deposited in an electroplating process onto the cathode. In electrorefining, the anodes consist of unrefined impure metal, and as the current passes through the acidic electrolyte the anodes are corroded into the solution so that the electroplating process deposits refined pure metal onto the cathodes.

History

Electrowinning is the oldest industrial electrolytic process. The English chemist Humphry Davy obtained sodium metal in elemental form for the first time in 1807 by the electrolysis of molten sodium hydroxide.

Electrorefining of copper was first demonstrated experimentally by Maximilian, Duke of Leuchtenberg in 1847.

James Elkington patented the commercial process in 1865 and opened the first successful plant in Pembrey, Wales in 1870. The first commercial plant in the United States was the Balbach and Sons Refining and Smelting Company in Newark, New Jersey in 1883.

Applications

The most common electrowon metals are lead, copper, gold, silver, zinc, aluminium, chromium, cobalt, manganese, and the rare-earth and alkali metals. For aluminium, this is the only production process employed. Several industrially important active metals (which react strongly with water) are produced commercially by electrolysis of their pyrochemical molten salts. Experiments using electrorefining to process spent nuclear fuel have been carried out. Electrorefining may be able to separate heavy metals such as plutonium, caesium, and strontium from the less-toxic bulk of uranium. Many electroextraction systems are also available to remove toxic (and sometimes valuable) metals from industrial waste streams.

Process

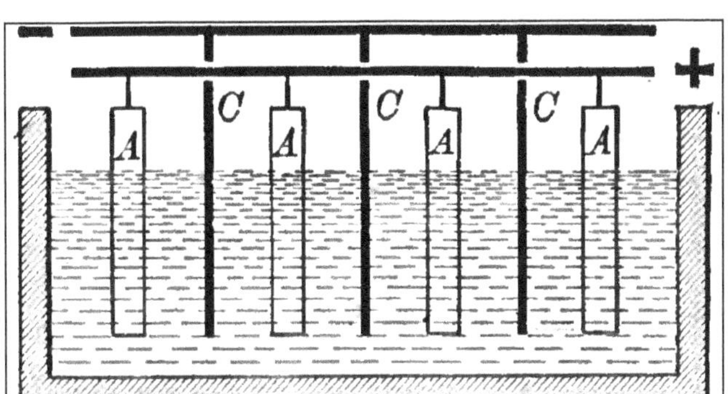

Apparatus for electrolytic refining of copper.

Most metals occur in nature in their oxidized form (ores) and thus must be reduced to their metallic forms. The ore is dissolved following some preprocessing in an aqueous electrolyte or in a molten salt and the resulting solution is electrolyzed. The metal is deposited on the cathode (either in solid or in liquid form), while the anodic reaction is usually oxygen evolution. Several metals are naturally present as metal sulfides; these include copper, lead, molybdenum, cadmium, nickel, silver, cobalt, and zinc. In addition, gold and platinum group metals are associated with sulfidic base metal ores. Most metal sulfides or their salts, are electrically conductive and this allows electrochemical redox reactions to efficiently occur in the molten state or in aqueous solutions.

Some metals, such as nickel do not electrolyze out but remain in the electrolyte solution. These are then reduced by chemical reactions to refine the metal. Other metals, which during the processing of the target metal have been reduced but not deposited at the cathode, sink to the bottom of the electrolytic cell, where they form a substance referred to as *anode sludge* or *anode slime*. The metals in this sludge can be removed by standard pyrorefining methods.

Because metal deposition rates are related to available surface area, maintaining properly working cathodes is important. Two cathode types exist, flat-plate and reticulated cathodes, each with its own advantages. Flat-plate cathodes can be cleaned and reused, and plated metals recovered. Reticulated cathodes have a much higher deposition rate compared to flat-plate cathodes. However, they are not reusable and must be sent off for recycling. Alternatively, starter cathodes of pre-refined metal can be used, which become an integral part of the finished metal ready for rolling or further processing.

Electroplating

Electroplating is a process that uses electric current to reduce dissolved metal cations so that they form a coherent metal coating on an electrode. The term is also used for electrical oxidation of anions onto a solid substrate, as in the formation silver chloride on silver wire to make silver/silver-chloride electrodes. Electroplating is primarily used to change the surface properties of an object (e.g. abrasion and wear resistance, corrosion protection, lubricity, aesthetic qualities, etc.), but may also be used to build up thickness on undersized parts or to form objects by electroforming.

Copper electroplating machine for layering PCBs.

The process used in electroplating is called electrodeposition. It is analogous to a galvanic cell acting in reverse. The part to be plated is the cathode of the circuit. In one technique, the anode is made of the metal to be plated on the part. Both components are immersed in a solution called an electrolyte containing one or more dissolved metal salts as well as other ions that permit the flow of electricity. A power supply supplies a direct current to the anode, oxidizing the metal atoms that it comprises and allowing them to dissolve in the solution. At the cathode, the dissolved metal ions in the electrolyte solution are reduced at the interface between the solution and the cathode, such that they "plate out" onto the cathode. The rate at which the anode is dissolved is equal to the rate at which the cathode is plated, vis-a-vis the current through the circuit. In this manner, the ions in the electrolyte bath are continuously replenished by the anode.

Other electroplating processes may use a non-consumable anode such as lead or carbon. In these techniques, ions of the metal to be plated must be periodically replenished in the bath as they are drawn out of the solution. The most common form of electroplating is used for creating coins such as pennies, which are small zinc plates covered in a layer of copper.

Process

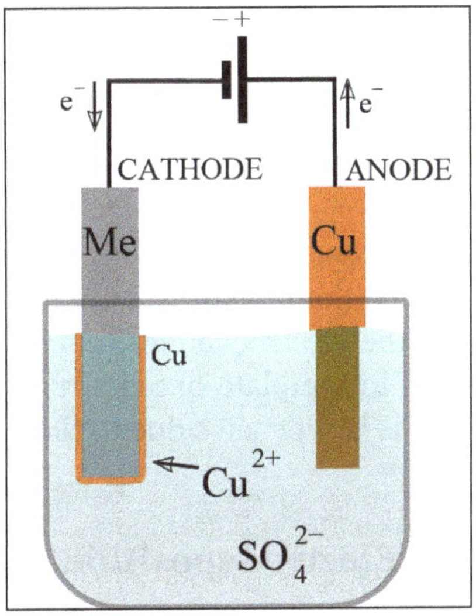

Electroplating of a metal (Me) with copper in a copper sulfate bath.

The cations associate with the anions in the solution. These cations are reduced at the cathode to deposit in the metallic, zero valence state. For example, in an acid solution, copper is oxidized at the anode to Cu^{2+} by losing two electrons. The Cu^{2+} associates with the anion SO_4^{2-} in the solution to form copper sulfate. At the cathode, the Cu^{2+} is reduced to metallic copper by gaining two electrons. The result is the effective transfer of copper from the anode source to a plate covering the cathode.

The plating is most commonly a single metallic element, not an alloy. However, some alloys can be electrodeposited, notably brass and solder.

Many plating baths include cyanides of other metals (e.g., potassium cyanide) in addition to cyanides of the metal to be deposited. These free cyanides facilitate anode corrosion, help to maintain a constant metal ion level and contribute to conductivity. Additionally, non-metal chemicals such as carbonates and phosphates may be added to increase conductivity.

When plating is not desired on certain areas of the substrate, stop-offs are applied to prevent the bath from coming in contact with the substrate. Typical stop-offs include tape, foil, lacquers, and waxes.

The ability of a plating to cover uniformly is called *throwing power*; the better the "throwing power" the more uniform the coating.

Strike

Initially, a special plating deposit called a "strike" or "flash" may be used to form a very thin (typically less than 0.1 micrometer thick) plating with high quality and good adherence to the substrate. This serves as a foundation for subsequent plating processes. A strike uses a high current density and a bath with a low ion concentration. The process is slow, so more efficient plating processes are used once the desired strike thickness is obtained.

The striking method is also used in combination with the plating of different metals. If it is desirable to plate one type of deposit onto a metal to improve corrosion resistance but this metal has inherently poor adhesion to the substrate, a strike can be first deposited that is compatible with both. One example of this situation is the poor adhesion of electrolytic nickel on zinc alloys, in which case a copper strike is used, which has good adherence to both.

Electrochemical Deposition

Electrochemical deposition is generally used for the growth of metals and conducting metal oxides because of the following advantages: (i) the thickness and morphology of the nanostructure can be precisely controlled by adjusting the electrochemical parameters, (ii) relatively uniform and compact deposits can be synthesized in template-based structures, (iii) higher deposition rates are obtained, and (iv) the equipment is inexpensive due to the non-requirements of either a high vacuum or a high reaction temperature.

Pulse Electroplating or Pulse Electrodeposition (PED)

A simple modification in the electroplating process is the pulse electroplating. This process involves the swift alternating of the potential or current between two different values resulting in a series of pulses of equal amplitude, duration and polarity, separated by zero current. By changing the pulse amplitude and width, it is possible to change the deposited film's composition and thickness.

Brush Electroplating

A closely related process is brush electroplating, in which localized areas or entire items are plated using a brush saturated with plating solution. The brush, typically a stainless steel body wrapped with a cloth material that both holds the plating solution and prevents direct contact with the item being plated, is connected to the positive side of a low voltage direct-current power source, and the item to be plated connected to the negative. The operator dips the brush in plating solution then applies it to the item, moving the brush continually to get an even distribution of the plating material. Brush electroplating has several advantages over tank plating, including portability, ability to plate items that for some reason cannot be tank plated (one application was the plating of portions of very large decorative support columns in a building restoration), low or no masking requirements, and comparatively low plating solution volume requirements. Disadvantages compared to tank plating can include greater operator involvement (tank plating can frequently be done with minimal attention), and inability to achieve as great a plate thickness.

Electroless Deposition

Usually an electrolytic cell (consisting of two electrodes, electrolyte, and external source of current) is used for electrodeposition. In contrast, an electroless deposition process uses only one electrode and no external source of electric current. However, the solution for the electroless process needs to contain a reducing agent so that the electrode reaction has the form:

$$M^{z+} + Red_{solution} \overset{\text{catalytic surface}}{\Rightarrow} M_{solid} + Oxy_{solution}$$

In principle any hydrogen-based reducer can be used although the redox potential of the reducer half-cell must be high enough to overcome the energy barriers inherent in liquid chemistry. Electroless nickel plating uses hypophosphite as the reducer while plating of other metals like silver, gold and copper typically use low molecular weight aldehydes.

A major benefit of this approach over electroplating is that the power sources and plating baths are not needed, reducing the cost of production. The technique can also plate diverse shapes and types of surface. The downside is that the plating process is usually slower and cannot create such thick plates of metal. As a consequence of these characteristics, electroless deposition is quite common in the decorative arts.

Cleanliness

Cleanliness is essential to successful electroplating, since molecular layers of oil can prevent adhesion of the coating. ASTM B322 is a standard guide for cleaning metals prior to electroplating. Cleaning processes include solvent cleaning, hot alkaline detergent cleaning, electro-cleaning, and acid treatment etc. The most common industrial test for cleanliness is the waterbreak test, in which the surface is thoroughly rinsed and held vertical. Hydrophobic contaminants such as oils cause the water to bead and break up, allowing the water to drain rapidly. Perfectly clean metal surfaces are hydrophilic and will retain an unbroken sheet of water that does not bead up or drain off. ASTM F22 describes a version of this test. This test does not detect hydrophilic contaminants, but the electroplating process can displace these easily since the solutions are water-based. Surfactants such as soap reduce the sensitivity of the test and must be thoroughly rinsed off.

Effects

Electroplating changes the chemical, physical, and mechanical properties of the workpiece. An example of a chemical change is when nickel plating improves corrosion resistance. An example of a physical change is a change in the outward appearance. An example of a mechanical change is a change in tensile strength or surface hardness which is a required attribute in tooling industry. Electroplating of acid gold on underlying copper/nickel-plated circuits reduces contact resistance as well as surface hardness. Copper-plated areas of mild steel act as a mask if case hardening of such areas are not desired. Tin-plated steel is chromium-plated to prevent dulling of the surface due to oxidation of tin.

History

Modern electrochemistry was invented by Italian chemist Luigi Valentino Brugnatelli (it) in 1805. Brugnatelli used his colleague Alessandro Volta's invention of five years earlier, the voltaic pile, to facilitate the first electrodeposition. Brugnatelli's inventions were suppressed by the French Academy of Sciences and did not become used in general industry for the following thirty years.

Luigu Valentino Brugnatelli.

Nickel plating.

By 1839, scientists in Britain and Russia had independently devised metal deposition processes similar to Brugnatelli's for the copper electroplating of printing press plates.

Boris Jacobi developed electroplating, electrotyping and galvanoplastic sculpture in Russia.

Boris Jacobi in Russia not only rediscovered galvanoplastics, but developed electrotyping and galvanoplastic sculpture. Galvanoplastics quickly came into fashion in Russia, with such people as inventor Peter Bagration, scientist Heinrich Lenz and science fiction author Vladimir Odoyevsky all contributing to further development of the technology. Among the most notorious cases of electroplating usage in mid-19th century Russia were gigantic galvanoplastic sculptures of St. Isaac's Cathedral in Saint Petersburg and gold-electroplated dome of the Cathedral of Christ the Saviour in Moscow, the tallest Orthodox church in the world.

Galvanoplastic sculpture on St. Isaac's Cathedral in Saint Petersburg.

The Woolrich Electrical Generator in Thinktank, Birmingham.

Soon after, John Wright of Birmingham, England discovered that potassium cyanide was a suitable electrolyte for gold and silver electroplating. Wright's associates, George Elkington and Henry Elkington were awarded the first patents for electroplating in 1840. These two then founded the electroplating industry in Birmingham from where it spread around the world. The Woolrich Electrical Generator of 1844, now in Thinktank, Birmingham Science Museum, is the earliest electrical generator used in an industrial process. It was used by Elkingtons.

The Norddeutsche Affinerie in Hamburg was the first modern electroplating plant starting its production in 1876.

As the science of electrochemistry grew, its relationship to the electroplating process became understood and other types of non-decorative metal electroplating processes were developed. Commercial electroplating of nickel, brass, tin, and zinc were developed by the 1850s. Electroplating baths and equipment based on the patents of the Elkingtons were scaled up to accommodate the plating of numerous large scale objects and for specific manufacturing and engineering applications.

The plating industry received a big boost with the advent of the development of electric generators in the late 19th century. With the higher currents available, metal machine components, hardware, and automotive parts requiring corrosion protection and enhanced wear properties, along with better appearance, could be processed in bulk.

The two World Wars and the growing aviation industry gave impetus to further developments and refinements including such processes as hard chromium plating, bronze alloy plating, sulfamate nickel plating, along with numerous other plating processes. Plating equipment evolved from manually operated tar-lined wooden tanks to automated equipment, capable of processing thousands of kilograms per hour of parts.

One of the American physicist Richard Feynman's first projects was to develop technology for electroplating metal onto plastic. Feynman developed the original idea of his friend into a successful invention, allowing his employer (and friend) to keep commercial promises he had made but could not have fulfilled otherwise.

Uses

Electroplating is widely used in various industries for coating metal objects with a thin layer of a different metal. The layer of metal deposited has some desired property, which the metal of the object lacks. For example, chromium plating is done on many objects such as car parts, bath taps, kitchen gas burners, wheel rims and many others for the fact that chromium is very corrosion resistant, and thus prolongs the life of the parts. Electroplating has wide usage in industries. It is also used in making inexpensive jewelry. Electroplating increases life of metal and prevents corrosion.

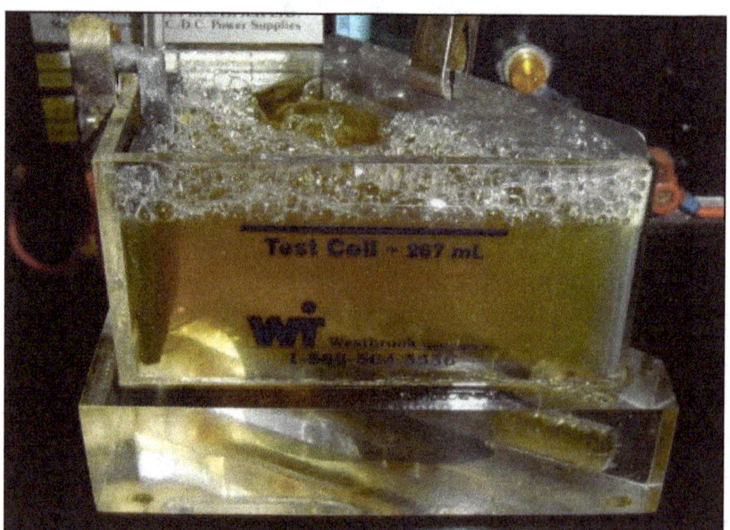

A zinc solution tested in a hull cell.

Hull Cell

The *Hull cell* is a type of test cell used to qualitatively check the condition of an electroplating bath. It allows for optimization for current density range, optimization of additive concentration, recognition of impurity effects and indication of macro-throwing power capability. The Hull cell replicates the plating bath on a lab scale. It is filled with a sample of the plating solution, an appropriate anode which is connected to a rectifier. The "work" is replaced with a hull cell test panel that will be plated to show the "health" of the bath.

The Hull cell is a trapezoidal container that holds 267 ml of solution. This shape allows one to place the test panel on an angle to the anode. As a result, the deposit is plated at different current densities which can be measured with a hull cell ruler. The solution volume allows for a quantitative optimization of additive concentration: 1 gram addition to 267 mL is equivalent to 0.5 oz/gal in the plating tank.

Haring-Blum Cell

The Haring-Blum Cell is used to determine the macro throwing power of a plating bath. The cell consists of 2 parallel cathodes with a fixed anode in the middle. The cathodes are at distances from the anode in the ratio of 1:5. The macro throwing power is calculating from the thickness of plating at the two cathodes when a dc current is passed for a specific period of time. The cell is fabricated out of Perspex or glass.

References

- Degarmo, E. Paul; Black, J T.; Kohser, Ronald A. (2003), Materials and Processes in Manufacturing (9th ed.), Wiley, p. 277, ISBN 0-471-65653-4
- Bard, Allan; Inzelt, Gyorgy; Scholz, Fritz, eds. (2012). "Haring-Blum Cell". Electrochemical Dictionary. Springer. p. 444. doi:10.1007/978-3-642-29551-5_8. ISBN 978-3-642-29551-5
- B. RAVI (1 January 2005). METAL CASTING: COMPUTER-AIDED DESIGN AND ANALYSIS. PHI Learning Pvt. Ltd. pp. 92–. ISBN 978-81-203-2726-9
- "Mining, Quarrying & Prospecting: The Difference between Mining, Quarrying & Prospecting". mqp-geotek.blogspot.co.uk. Retrieved 02, March 2020
- Ott, J. Bevan; Boerio-Goates, Juliana (2000), Chemical Thermodynamics: Advanced Applications, Academic Press, pp. 92–93, ISBN 0-12-530985-6
- Jane L. Bassett; Peggy Fogelman; David A. Scott; Ronald C. Schmidtling (2008). The Craftsman Revealed: Adriaen de Vries, Sculptor in Bronze. Getty Publications. pp. 269–. ISBN 978-0-89236-919-5
- Atkins, Peter; Jones, Loretta (2008), Chemical Principles: The Quest for Insight (4th ed.), W. H. Freeman and Company, p. 236, ISBN 0-7167-7355-4
- T F Waters (11 September 2002). Fundamentals of Manufacturing For Engineers. CRC Press. pp. 17–. ISBN 978-0-203-50018-7
- "Assessing the opportunities of landfill mining - Research database - University of Groningen". www.rug.nl. Retrieved 21, May 2020
- Gordon Elliott (2006). Aspects of Ceramic History: A Series of Papers Focusing on the Ceramic Artifact As Evidence of Cultural and Technical Developments. Gordon Elliott. pp. 52–. ISBN 978-0-9557690-0-9

Permissions

All chapters in this book are published with permission under the Creative Commons Attribution Share Alike License or equivalent. Every chapter published in this book has been scrutinized by our experts. Their significance has been extensively debated. The topics covered herein carry significant information for a comprehensive understanding. They may even be implemented as practical applications or may be referred to as a beginning point for further studies.

We would like to thank the editorial team for lending their expertise to make the book truly unique. They have played a crucial role in the development of this book. Without their invaluable contributions this book wouldn't have been possible. They have made vital efforts to compile up to date information on the varied aspects of this subject to make this book a valuable addition to the collection of many professionals and students.

This book was conceptualized with the vision of imparting up-to-date and integrated information in this field. To ensure the same, a matchless editorial board was set up. Every individual on the board went through rigorous rounds of assessment to prove their worth. After which they invested a large part of their time researching and compiling the most relevant data for our readers.

The editorial board has been involved in producing this book since its inception. They have spent rigorous hours researching and exploring the diverse topics which have resulted in the successful publishing of this book. They have passed on their knowledge of decades through this book. To expedite this challenging task, the publisher supported the team at every step. A small team of assistant editors was also appointed to further simplify the editing procedure and attain best results for the readers.

Apart from the editorial board, the designing team has also invested a significant amount of their time in understanding the subject and creating the most relevant covers. They scrutinized every image to scout for the most suitable representation of the subject and create an appropriate cover for the book.

The publishing team has been an ardent support to the editorial, designing and production team. Their endless efforts to recruit the best for this project, has resulted in the accomplishment of this book. They are a veteran in the field of academics and their pool of knowledge is as vast as their experience in printing. Their expertise and guidance has proved useful at every step. Their uncompromising quality standards have made this book an exceptional effort. Their encouragement from time to time has been an inspiration for everyone.

The publisher and the editorial board hope that this book will prove to be a valuable piece of knowledge for students, practitioners and scholars across the globe.

Index

www.ingramcontent.com/pod-product-compliance
Lightning Source LLC
LaVergne TN
LVHW080904210926
848485LV00004B/586